Tropical Community Tree Guide: Benefits, Costs, and Strategic Planting

Kelaine E. Vargas, E. Gregory McPherson, James R. Simpson, Paula J. Peper, Shelley L. Gardner, and Qingfu Xiao

U.S. Department of Agriculture, Forest Service

Pacific Southwest Research Station

Albany, California

General Technical Report PSW-GTR-216

September 2008

This work was sponsored by the U.S. Department of Agriculture, Forest Service, State and Private Forestry, Urban and Community Forestry Program; the Kaulunani Urban Forestry Program of the Hawaii Department of Land and Natural Resources, Division of Forestry and Wildlife; and the City and County of Honolulu, Department of Parks and Recreation, Urban Forestry Division.

Abstract

Vargas, Kelaine E.; McPherson, E. Gregory; Simpson, James R.; Peper, Paula J.; Gardner, Shelley L.; Xiao, Qingfu. 2008. Tropical community tree guide: benefits, costs, and strategic planting. Gen. Tech. Rep. PSW-GTR-216. Albany, CA: U.S. Department of Agriculture, Forest Service, Pacific Southwest Research Station. 109 p.

Even as they increase the beauty of our surroundings, trees provide us with a great many ecosystem services, including air quality improvement, energy conservation, stormwater interception, and atmospheric carbon dioxide reduction. These benefits must be weighed against the costs of maintaining trees, including planting, pruning, irrigation, administration, pest control, liability, cleanup, and removal. We present benefits and costs for representative small, medium, and large trees in the Tropical region derived from models based on indepth research carried out in Honolulu, Hawaii. Average annual net benefits increase with tree size and differ based on location: $9 (public) to $30 (yard) for a small tree, $43 (public) to $79 (yard) for a medium tree, $70 (public) to $92 (yard) for a large tree. Two hypothetical examples of planting projects are described to illustrate how the data in this guide can be adapted to local uses, and guidelines for maximizing benefits and reducing costs are given.

Keywords: Ecosystem services, Tropical region, urban forestry, benefit-cost analysis.

In the Tropical region, trees play an environmental, cultural, and historical role in communities. Here a grand old monkeypod tree graces the grounds of Hawaii's Iolani Palace, the home of the Hawaiian Kingdom's last two monarchs.

Summary

Trees provide many valuable ecosystem services: they reduce energy consumption, they trap and filter stormwater, they help clean the air by intercepting air pollutants, and they help in the fight against global climate change by sequestering carbon dioxide (CO_2). At the same time, they provide a wide array of aesthetic, social, economic, and health benefits that are less tangible.

This report quantifies benefits and costs for representative small, medium, and large trees in the Tropical region: the species chosen as representative are the silver buttonwood, rainbow shower tree, and monkeypod (see "Common and Scientific Names" section). The analysis describes "yard trees" (those planted in residential sites) and "public trees" (those planted on streets or in parks). Benefits are calculated based on tree growth curves and numerical models that consider regional climate, building characteristics, air pollutant concentrations, and prices. Tree care costs and mortality rates are based on results from a survey of municipal and commercial arborists. We assume a 19 percent mortality rate over a 40-year timeframe.

The measurements used in modeling environmental and other benefits of trees are based on indepth research carried out in Honolulu, Hawaii. Given the Tropical region's diverse geographical area, this approach provides general approximations based on some necessary assumptions that serve as a starting point for more specific local calculations. It is a general accounting that can be easily adapted and adjusted for local planting projects. Two examples are provided that illustrate how to adjust benefits and costs to reflect different aspects of local urban forest improvement projects.

Large trees provide the most benefits. Average annual benefits increase with mature tree size and differ based on tree location. The lowest values are for public trees or yard trees on the southern side of houses, and the highest values are for yard trees on the western side of houses. Average annual benefits range as follows:

- $33 to $42 for a small tree
- $72 to $92 for a medium tree
- $105 to $108 for a large tree

Benefits associated with reduced energy use and increased aesthetic and other benefits reflected in higher property values account for the largest proportion of total benefits in this region. Reduced levels of stormwater runoff, air pollutants, and CO_2 in the air are the next most important benefits.

Energy conservation benefits differ with tree location as well as size. Trees located opposite west-facing walls provide the greatest cooling energy savings. Reducing energy needs reduces CO_2 emissions and thereby reduces atmospheric CO_2. Similarly, energy savings that reduce demand from powerplants account for important reductions in gases that produce ozone, a major component of smog, and other air pollutants.

The benefits of trees are offset by the costs of caring for them. Based on our surveys of municipal and residential arborists, the average annual cost for tree care ranges from $14 to $33 per tree. (Values below are for yard and public trees, respectively.)

- $12 and $24 for a small tree
- $14 and $29 for a medium tree
- $16 and $35 for a large tree

Planting costs, annualized over 40 years, are the greatest expense for yard trees ($7.50 per tree per year); planting costs for public trees are significantly lower ($3 per tree per year). For public trees, pruning ($10 to $17 per tree per year) and managing conflicts with infrastructure ($5 to $8 per tree per year) are the greatest costs. Public trees also incur administrative expense ($2 to $3 per tree per year).

Average annual net benefits (benefits minus costs) per tree for a 40-year period differ with tree location and tree size and range from a low of $9 to a high of $92 per tree.

- $9 for a small public tree to $30 for a small yard tree on the west side of a house
- $43 for a medium public tree to $79 for a medium yard tree on the west side of a house
- $70 for a large public tree to $92 for a large yard tree on the west side of a house

Environmental benefits alone, including energy savings, stormwater runoff reduction, improved air quality, and reduced atmospheric CO_2, are up to five times tree care costs.

Net benefits are substantial when summed over the 40-year period (values below are for yard trees opposite a west wall and public trees, respectively):

- $1,330 and $340 for a small tree
- $3,535 and $1,905 for a medium tree
- $4,115 and $3,060 for a large tree

Yard trees produce higher net benefits than public trees, primarily because of lower maintenance costs.

To demonstrate ways that communities can adapt the information in this report to their needs, examples of two fictional cities interested in improving their urban forest have been created. The benefits and costs of different planting projects are determined. In the hypothetical city of Punakea Valley, net benefits and benefit-cost ratios (BCRs; total benefits divided by costs) are calculated for a planting of 1,000 trees (2-in caliper) assuming a cost of $200 per tree, 19 percent mortality rate, and 40-year analysis. Total benefits are $3.9 million, total costs are about $1.3 million, and net benefits are $2.6 million ($65.31 per tree per year). The BCR is 3.00:1, indicating that $3.00 is returned for every $1 invested. The net benefits and BCRs (in parentheses) by mature tree size are:

- $73,739 (1.47:1) for 150 silver buttonwood trees
- $718,734 (2.65:1) for 350 rainbow shower trees
- $1.8 million (3.54:1) for 500 monkeypods

Increased property values reflecting aesthetic and other benefits of trees (48.5 percent) made up the largest share, and reduced energy accounted for another 30 percent. Reduced stormwater runoff (16 percent), air quality improvement (4.5 percent), and atmospheric CO_2 reduction (1 percent) make up the remaining benefits.

In the fictional city of Mangrovia, long-term planting and tree care costs and benefits were compared to determine if a current fashion for planting small flowering trees instead of the large stately trees that were once standard is substantially affecting the level of benefits residents are receiving. Over a 40-year period, the net benefits are:

- $313 for a small tree
- $1,558 for a medium tree
- $2,448 for a large tree

Based on this analysis, the city of Mangrovia decided to strengthen its tree ordinance, requiring developers to plant large trees whenever feasible and to create tree shade plans that show how they will achieve 50 percent shade over streets, sidewalks, and parking lots within 15 years of development.

Contents

Chapter 1. Introduction

The Tropical Region

From the Hawaiian islands to Puerto Rico and the very tip of southern Florida, the Tropical region (fig.1) contains a diverse assemblage of municipalities and environments. Tiny communities in the Florida Keys contrast with major cities, such as Honolulu and San Juan, and the rugged, mountainous terrain of Hawaii's Big Island is a counterpoint to the Florida coast, which sits barely above sea level. The different layers of history, the many cultures, and the languages that characterize this region add to its diversity.

Figure 1—The Tropical region covers the Hawaiian and other Pacific Islands, Puerto Rico and the Caribbean, and the very southern tip of Florida and the Florida Keys. Reference cities for the Tropical and other nearby regions are indicated with large circles.

The **climate**[1] of this region, which corresponds to Sunset climate zone 25 (Brenzel 2001) and USDA hardiness zones 10 and 11, is characterized by high humidity, warm to hot temperatures throughout the year, and no periods of frost. Average high temperatures range from the high 80s in July to the mid 70s in January. On the islands of the Tropical region, precipitation rates differ widely even across small areas. Across the island of Oahu, Hawaii, rainfall ranges from as little as 17 in near Honolulu to as much as 160 inches in the mountains. The same is true in Puerto Rico, where the southern part of the island receives about 40 in of rain each year and the rain forests about 180 in. Across southern Florida, rainfall is more regular, with annual averages between 60 and 65 in on the mainland and around 35 inches in the Florida Keys.

As the communities of the Tropical region continue to grow and change during the coming decades, growing and sustaining healthy **community forests** is integral to the quality of life that residents experience. The urban forest is a distinctive feature of the landscape that protects us from the elements, cleans the water we drink and the air we breathe, and forms a connection to earlier generations who planted and tended the trees.

The role of urban forests in enhancing the environment, increasing community attractiveness and livability, and fostering civic pride takes on greater significance as communities strive to balance economic growth with environmental quality and social well-being. The simple act of planting trees provides opportunities to connect residents with nature and with each other (fig. 2). Neighborhood tree plantings and stewardship projects stimulate investment by local citizens, businesses, and governments for the betterment of their communities. Community forests bring opportunity for economic renewal, combating development woes, and increasing the quality of life for community residents.

Tropical communities can promote energy efficiency through tree planting and stewardship programs that strategically locate trees to save energy and minimize conflicts with urban infrastructure. The same trees can provide additional benefits by reducing stormwater runoff; improving local air, soil, and water quality; reducing atmospheric carbon dioxide (CO_2); providing wildlife habitat; increasing property values; slowing traffic; enhancing community attractiveness and investment; and promoting human well-being.

---

[1] Words in bold are defined in the glossary.

Figure 2—Tree planting and stewardship programs provide opportunities for local residents to work together to build better communities.

This guide builds upon studies by the USDA Forest Service in Chicago and Sacramento (McPherson et al. 1994, 1997), and other regional tree guides from the Center for Urban Forest Research (McPherson et al. 1999b, 2000, 2003, 2004, 2006a, 2006b, 2006c, 2007, Vargas et al. 2007b, 2007c) to extend knowledge of urban forest benefits in the Tropical region. The guide:

- Quantifies benefits of trees on a per-tree basis rather than on a canopy cover basis (it should not be used to estimate benefits for trees growing in forest stands).
- Describes management costs and benefits.
- Details how tree planting programs can improve environmental quality, conserve energy, and add value to communities.
- Explains where to place residential yard and public trees to maximize their benefits and cost-effectiveness.
- Describes ways to minimize conflicts between trees and power lines, sidewalks, and buildings.
- Illustrates how to use this information to estimate benefits and costs for local tree planting projects.

These guidelines are specific to the Tropical region and are based on data and calculations from open-growing urban trees in this region.

Street, park, and shade trees are components of all Tropical communities, and they affect every resident. Their benefits are myriad. However, with municipal tree programs dependent on taxpayer-supported general funds, communities are forced to ask whether trees are worth the price to plant and care for over the long term, thus requiring urban forestry programs to demonstrate their cost-effectiveness (McPherson 1995). If tree plantings are proven to benefit communities, then financial commitment to tree programs will be justified. Therefore, the objective of this tree guide is to identify and describe the benefits and costs of planting trees in Tropical communities—providing a tool for municipal tree managers, arborists, and tree enthusiasts to increase public awareness and support for trees (Dwyer and Miller 1999).

Trees in tropical communities enhance quality of life.

Chapter 2. Benefits and Costs of Urban and Community Forests

This chapter describes benefits and costs of public and privately managed trees and presents the functional benefits and associated economic value of community forests. Expenditures related to tree care and management are assessed—a necessary process for creating cost-effective programs (Dwyer et al. 1992, Hudson 1983).

Benefits

Saving Energy

Energy is an essential ingredient for quality of life and for economic growth. Conserving energy by greening our cities is often more cost-effective than building new powerplants. For example, while California was experiencing energy shortages in 2001, its 177 million city trees were providing shade and conserving energy. Annual savings to utilities were an estimated $500 million in wholesale electricity and generation purchases (McPherson and Simpson 2003). Planting 50 million more shade trees in strategic locations would provide savings equivalent to seven 100-megawatt powerplants. The cost of reducing the peak load was $63 per kW, considerably less than the $150 per kW threshold that is considered cost-effective. Utility companies in the Tropical region and throughout the country can invest in shade tree programs as a cost-effective energy conservation measure to lower peak energy demands.

Trees in the Tropical region modify climate and conserve building energy use in two principal ways (fig. 3):

- Shading reduces the amount of heat absorbed and stored by built surfaces, including buildings and paved areas.
- Evapotranspiration converts liquid water to water vapor and thus cools the air by using solar energy that would otherwise result in heating of the air.

Summer temperatures in cities can be 3 to 8 °F warmer than temperatures in the surrounding countryside. This is known as the **urban heat island** effect. Trees and other vegetation can combat this warming effect at small and large scales. On individual building sites, trees may lower air temperatures up to 5 °F compared with outside the **greenspace**. At larger scales (6 mi^2), temperature differences of more than 9 °F have been observed between city centers and more vegetated suburban areas (Akbari et al. 1992). A recent study by scientists at the National

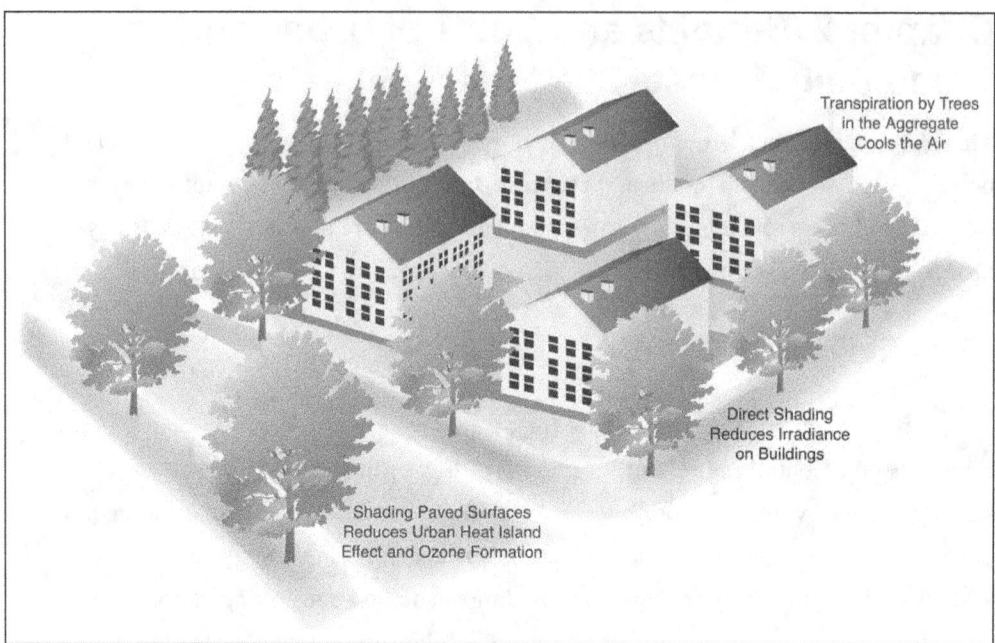

Figure 3—Trees save energy for cooling by shading buildings and lowering summertime temperatures (drawing by Mike Thomas).

Aeronautics and Space Administration and Columbia University found that street trees provide the "greatest cooling potential per unit area" (Rosenzweig et al. 2006).

For individual buildings, strategically placed trees can increase energy efficiency. Because the Sun is low in the east and west for several hours each day, trees that shade these walls in particular will help keep buildings cool.

Trees provide greater energy savings in the Tropical region than in milder climate regions because they can have cooling effects year round. In Miami, for example, trees were found to produce substantial cooling savings for an energy-efficient two-story wood-frame house (McPherson et al. 1993). A typical energy-efficient house with air conditioning requires about $546 each year for cooling. A computer simulation demonstrated that three 25-ft tall trees—two on the west side of the house and one on the east—would save $150 each year for cooling, a 28-percent reduction.

A recent study of some of the municipal trees of Honolulu showed that the 43,817 trees save city residents approximately $343,356 in annual air conditioning (Vargas et al. 2007a) or $7.84 per tree. The largest trees provide the largest benefits; the large monkeypod (see "Common and Scientific Names" section), for example, accounted for 10.6 percent of the energy benefits although it represented only 3.1 percent of the population.

Conserving energy by greening our cities is important because it can be more cost-effective than building new powerplants (for more information, see the Center for Urban Forest Research's research summaries *Green Plants or Power Plants?* [Geiger 2001] and *Save Dollars With Shade* [Geiger 2002a]). In the Tropical region, there is ample opportunity to "retrofit" communities with more sustainable landscapes through strategic tree planting and care of existing trees.

Reducing Atmospheric Carbon Dioxide

Global temperatures have increased since the late 19th century, with major warming periods from 1910 to 1945 and from 1976 to the present (IPCC 2001). Human activities, primarily fossil-fuel consumption, are adding greenhouse gases to the atmosphere, and current research suggests that the recent increases in temperature can be attributed in large part to increases in greenhouse gases (IPCC 2001). Higher global temperatures are expected to have a number of adverse effects, including raising ocean levels, an aspect of particular concern in parts of the Tropical region (Meehl et al. 2005). Increasing frequency of extreme weather events will continue to tax emergency management resources.

Urban forests have been recognized as important storage sites for carbon dioxide (CO_2), the primary greenhouse gas (Nowak and Crane 2002). Private markets dedicated to reducing CO_2 emissions by trading carbon credits are emerging (Chicago Climate Exchange 2007, McHale et al. 2007). Carbon credits have sold for as much as EUR 33 per ton (~$40; European Climate Exchange 2006), and the social costs of CO_2 emissions (an estimate of the monetary value of worldwide damage done by anthropogenic CO_2 emissions) are estimated to range from £4 to £27 per ton ($7 to $47 per ton) (Pearce 2003). For comparison, for every $20 spent on a tree planting project in Arizona, 1 ton of atmospheric CO_2 was reduced (McPherson and Simpson 1999). As carbon trading markets become accredited and prices rise, these markets could provide monetary resources for community forestry programs.

Urban forests can reduce atmospheric CO_2 in two ways (fig. 4):

- Trees directly sequester CO_2 in their stems, leaves, and roots while they grow.
- Trees near buildings can reduce the demand for air conditioning, thereby reducing emissions associated with power production.

At the same time, the positive impact of trees on CO_2 is offset by some emissions. To provide a complete picture of atmospheric CO_2 reductions from tree

Figure 4—Trees sequester carbon dioxide (CO_2) as they grow and indirectly reduce CO_2 emissions from powerplants through energy conservation. At the same time, CO_2 is released through decomposition and tree care activities that involve fossil-fuel consumption (drawing by Mike Thomas).

plantings, it is important to consider CO_2 released into the atmosphere through tree planting and care activities, as well as decomposition of wood from pruned or dead trees. During the process of planting and maintaining trees, vehicles, chain saws, chippers, and other equipment release CO_2. Typically, CO_2 released from tree planting, maintenance, and other tree-related activities is about 2 to 8 percent of annual CO_2 reductions obtained through sequestration and **reduced powerplant emissions** (McPherson and Simpson 1999). And eventually, all trees die, and most of the carbon that has accumulated in their structure is released into the atmosphere as CO_2 through decomposition. The rate of release into the atmosphere depends on if and how the wood is reused. For instance, recycling of urban wood waste into products such as furniture can delay the rate of decomposition compared to its reuse as mulch. Tree waste can also be used as a fuel source to generate electricity. If this biomass fuel replaces a more carbon-intensive form of electricity production, there will be an overall reduction in atmospheric CO_2.

Regional variations in climate and the mix of fuels that produce energy to cool buildings influence potential CO_2 emission reductions. The average emission rate in Honolulu, Hawaii, is 1,854 lb of CO_2 per MWh (US EPA 2006b), a high value, because 99.7 percent of Honolulu's power is generated from oil. The state of Florida, on the other hand, derives its energy from less CO_2-intensive sources—a mix of coal, oil, natural gas, and nuclear power—and therefore has an average emission of 1,327 lbs of CO_2 per MWh, which is close to the national average of 1,363 lbs of CO_2 per MWh (US EPA 2006b). Cities in the Tropical region with relatively high CO_2 emission rates will see greater benefits from reduced energy demand relative to other areas with lower emissions rates.

A study of the municipal trees of Honolulu found that the 43,817 trees in the inventory sequester about 1,683 tons of CO_2 (Vargas et al. 2007a) annually and, by reducing energy use, reduce the production of CO_2 at the powerplant by 1,796 tons. Approximately 139 tons of CO_2 is released from decaying trees and during maintenance, with a positive net reduction in CO_2 from trees of 3,340 tons.

Another study in Chicago focused on the carbon sequestration benefit of residential tree canopy. Tree **canopy cover** in two residential neighborhoods was estimated to sequester on average 0.112 lb/ft^2, and pruning activities released 0.016 lb/ft^2 (Jo and McPherson 1995). Net annual carbon uptake was 0.096 lb/ft^2.

Grass-roots tree-planting efforts to reduce atmospheric CO_2 can be very successful. Since 1990, Trees Forever, an Iowa-based nonprofit organization, has planted trees for energy savings and atmospheric CO_2 reduction with utility sponsorships. Over 1 million trees have been planted in 400 communities with the help of 120,000 volunteers. These trees are estimated to offset CO_2 emissions by 50,000 tons annually. Survival rates are an amazing 91 percent, indicating a highly trained and committed volunteer force (Thompson et al. 2004).

Improving Air Quality

Approximately 159 million people live in areas where **ozone** (O_3) concentrations violate federal air quality standards. About 100 million people live in areas where dust and other small particulate matter (PM_{10}) exceed levels for healthy air. Air pollution is a serious health threat to many city dwellers, contributing to asthma, coughing, headaches, respiratory and heart disease, and cancer (Smith 1990). Impaired health results in increased social costs for medical care, greater absentee-ism, and reduced longevity.

Recently, the Environmental Protection Agency (EPA) recognized tree planting as a measure in state implementation plans for reducing O_3. Air quality management districts have funded tree planting projects to control particulate matter. These policy decisions are creating new opportunities to plant and care for trees as a method for controlling air pollution (Luley and Bond 2002; for more information see www.treescleanair.org [USDA FS 2006b] and the Center for Urban Forest Research's research summary *Trees—The Air Pollution Solution* [Geiger 2006]).

Urban forests provide a number of air quality benefits (fig. 5):

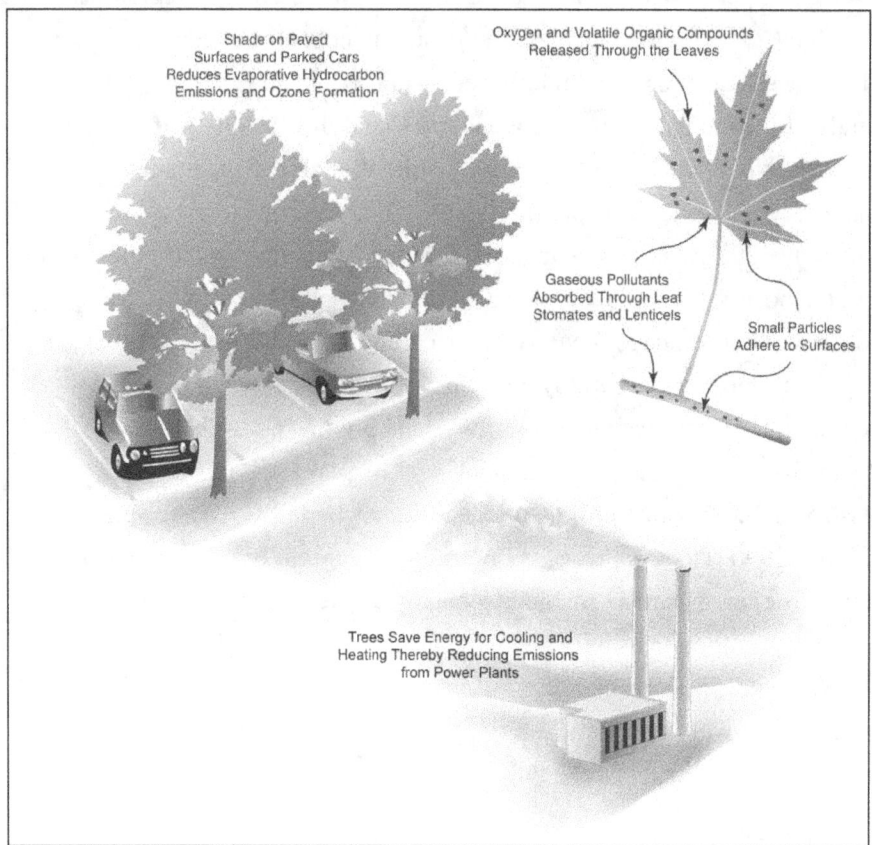

Figure 5—Trees absorb gaseous pollutants, retain particles on their surfaces, and release oxygen and volatile organic compounds. By cooling urban heat islands and shading parked cars, trees can reduce ozone formation (drawing by Mike Thomas).

- They absorb gaseous pollutants (e.g., O_3, nitrogen dioxide [NO_2], and sulfur dioxide [SO_2]) through leaf surfaces.
- They intercept PM_{10} (e.g., dust, ash, pollen, smoke).
- They release oxygen through photosynthesis.
- They reduce energy use, which reduces emissions of pollutants from powerplants, including NO_2, SO_2, PM_{10}, and volatile organic compounds (**VOCs**).
- They shade paved surfaces and parked cars, which lowers air temperatures, reducing hydrocarbon emissions and O_3 levels.

Trees may also adversely affect air quality. Most trees emit **biogenic volatile organic compounds** (BVOCs) such as isoprenes and monoterpenes that can contribute to O_3 formation. The contribution of BVOC emissions from city trees to O_3 formation depends on complex geographic and atmospheric interactions that have not been studied in most cities. Some complicating factors include variations with temperature and atmospheric levels of NO_2.

A computer simulation study for Atlanta suggested that it would be very difficult to meet EPA ozone standards in the region by using trees because of the high BVOC emissions from native pines and other vegetation (Chameides et al. 1988). The results, however, were not straightforward. A later study showed that although removing trees reduced BVOC emissions, any positive effect was overwhelmed by increased hydrocarbon emissions from natural and anthropogenic sources owing to the increased air temperatures associated with tree removal (Cardelino and Chameides 1990). A similar finding was reported for the Houston-Galveston area, where deforestation associated with urbanization from 1992 to 2000 increased surface temperatures. Despite the decrease in BVOC emissions, O_3 concentrations increased because of the enhanced urban heat island effect during simulated episodes (Kim et al. 2005).

As well, the O_3-forming potential of different tree species differs considerably (Benjamin and Winer 1998). In a study in the Los Angeles basin, increased planting of low-BVOC-emitting tree species was shown to reduce O_3 concentrations, whereas planting of medium and high emitters would increase overall O_3 concentrations (Taha 1996). A study in the Northeastern United States, however, found that species mix had no detectable effects on O_3 concentrations (Nowak et al. 2000). Although new trees increased BVOC emissions, ambient VOC emissions were so high that additional BVOCs had little effect on air quality. These potentially negative effects of trees on one kind of air pollution must be considered in light of their great benefit in other areas.

Trees absorb gaseous pollutants through stomates, tiny openings in the leaves. Other methods of pollutant removal include adsorption of gases to plant surfaces and uptake through bark pores. Once gases enter the leaf, they diffuse into intercellular spaces, where some react with inner leaf surfaces and others are absorbed by water films to form acids. Pollutants can damage plants by altering their metabolism and growth. At high concentrations, pollutants cause visible damage to leaves, such as spotting and bleaching (Costello and Jones 2003). Although some pollutants may pose health hazards to plants, pollutants such as nitrogenous gases can also be sources of essential nutrients for them.

Trees intercept small airborne particles. Some particles that are intercepted by a tree are absorbed, but most adhere to plant surfaces. Species with hairy or rough leaf, twig, and bark surfaces are efficient interceptors (Smith and Dochinger 1976). Intercepted particles are often resuspended to the atmosphere when wind blows the branches, and rain will wash some particulates off plant surfaces. The ultimate fate of these pollutants depends on whether they fall onto paved surfaces and enter the stormwater system, or fall on pervious surfaces, where they are filtered in the soil.

Trees near buildings can reduce the demand for air conditioning, thereby reducing emissions of PM_{10}, SO_2, NO_2, and VOCs associated with electric power production, an effect that can be sizable. For example, a strategically located tree can save 100 kWh in electricity for cooling annually (McPherson and Simpson 1999, 2002, 2003). Assuming that this conserved electricity comes from a typical new coal-fired powerplant in the Tropical region, the tree reduces emissions of SO_2 by 0.35 lb, NO_2 by 0.26 lb (US EPA 2006b), and PM_{10} by 0.84 lb (US EPA 1998). The same tree is responsible for conserving 60 gal of water in cooling towers and reducing CO_2 emissions by 172 lb.

A study of a part of the urban forest of Honolulu found that the 43,817 trees in the city's inventory removed 9 tons of air pollutants, an environmental service valued at $47,365 (Vargas et al. 2007a). The forest within the urbanized area of Jacksonville, Florida, (approximately 40,000 acres) was estimated to remove 3.8 million tons of air pollutants, a service valued at $9.4 million (American Forests 2005). A more recent study in Palm Beach County, Florida, assessed the damage done to the urban canopy by recent hurricanes and calculated the benefits lost. Between 2004 and 2006, the tree canopy of the urbanized parts of the county declined by 38 percent and thereby increased the level of air pollutants in the atmosphere by approximately 2.3 million pounds (American Forests 2007).

Trees in a Davis, California, parking lot were found to improve air quality by reducing air temperatures 1 to 3 °F (Scott et al. 1999). By shading asphalt surfaces and parked vehicles, trees reduce hydrocarbon emissions (VOCs) from gasoline that evaporates out of leaky fuel tanks and worn hoses (for more information, see our research summary *Where Are All the Cool Parking Lots?* [Geiger 2002b]). These evaporative emissions are a principal component of smog, and parked vehicles are a primary source (fig. 6). In California, parking lot tree plantings can be funded as an air quality improvement measure because of the associated reductions in evaporative emissions.

Figure 6—Trees planted to shade parking areas can reduce hydrocarbon emissions and improve air quality.

Reducing Stormwater Runoff and Improving Hydrology

Urban stormwater runoff is a major source of pollution entering wetlands, streams, lakes, and oceans. Healthy trees can reduce the amount of runoff and pollutants in receiving waters (Cappiella et al. 2005). This is important because federal law requires states and localities to control nonpoint-source pollution, such as runoff from pavements, buildings, and landscapes. Trees are mini-reservoirs, controlling runoff at the source, thereby reducing runoff volumes and erosion of watercourses, as well as delaying the onset of **peak flows**. Trees can reduce runoff in several ways (fig. 7; for more information, see our research summary *Is All Your Rain Going Down the Drain?* [Geiger 2003]):

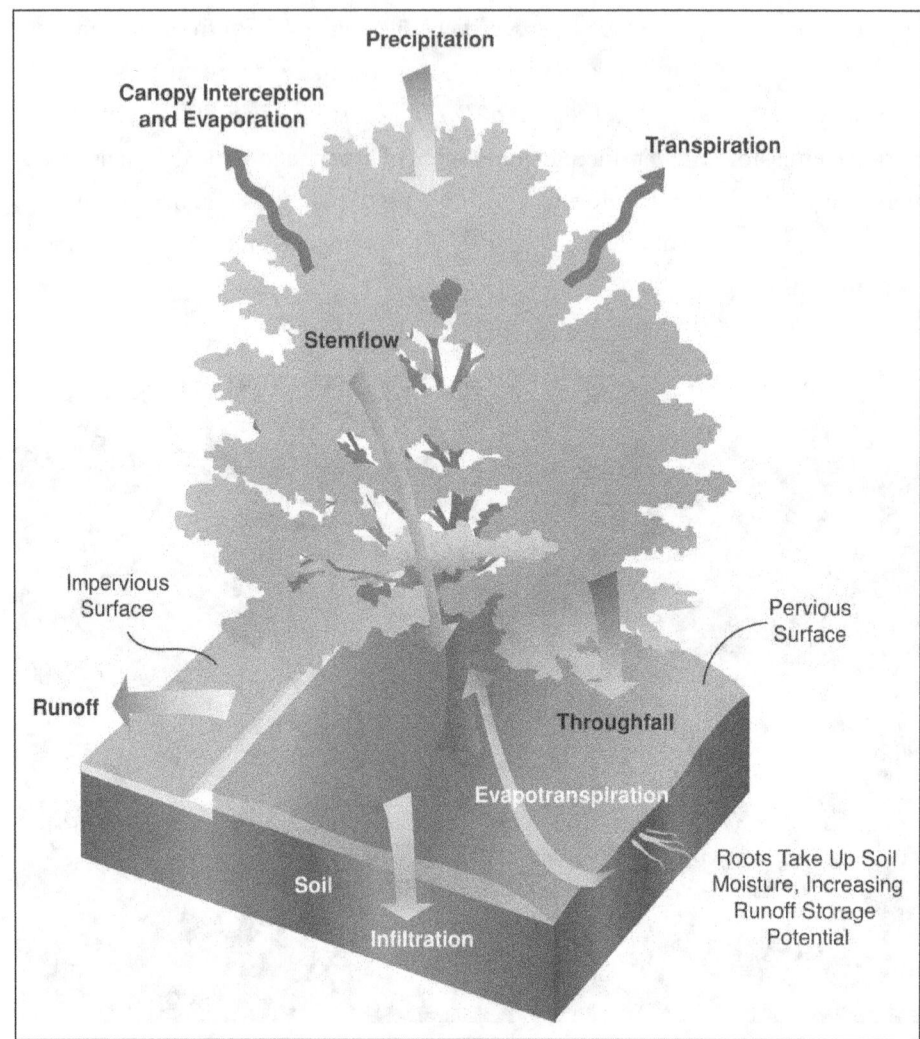

Figure 7—Trees intercept a portion of rainfall that evaporates and never reaches the ground. Some rainfall runs to the ground along branches and stems (stemflow) and some falls through gaps or drips off leaves and branches (throughfall). Transpiration increases soil moisture storage potential (drawing by Mike Thomas).

- Leaves and branch surfaces intercept and store rainfall, thereby reducing runoff volumes and delaying the onset of peak flows.
- Roots reduce soil compaction, increasing the rate at which rainfall infiltrates soil and the capacity of soil to store water, reducing overland flow.
- Tree canopies reduce soil erosion by diminishing the impact of raindrops on barren surfaces.
- **Transpiration** through tree leaves reduces moisture levels in the soil, increasing the soil's capacity to store rainfall.

Rainfall that is stored temporarily on canopy leaf and bark surfaces is called intercepted rainfall. Intercepted water evaporates, drips from leaf surfaces, and flows down stem surfaces to the ground. Tree surface saturation generally occurs after 1 to 2 in of rain has fallen (Xiao et al. 2000). During large storm events, rainfall exceeds the amount that the tree crown can store, about 50 to 100 gal per tree. The interception benefit is the amount of rainfall that does not reach the ground because it evaporates from the crown. As a result, the volume of runoff is reduced and the time of peak flow is delayed. Trees protect water quality by substantially reducing runoff during small rainfall events that are responsible for most pollutant washoff. Therefore, urban forests generally produce more benefits through water quality protection than through flood control (Xiao et al. 1998, 2000).

The amount of rainfall trees intercept depends on their architecture, rainfall patterns, and climate. Tree-crown characteristics that influence interception are the trunk, stem, surface areas, textures, area of gaps, period when leaves are present, and dimensions (e.g., tree height and diameter). Trees with coarse surfaces retain more rainfall than those with smooth surfaces. Large trees generally intercept more rainfall than small trees do because greater surface areas allow for greater evaporation rates. Tree crowns with few gaps reduce **throughfall** to the ground. Species that are in leaf when rainfall is plentiful are more effective than deciduous species that have dropped their leaves during the rainy season.

Studies that have simulated urban forest effects on stormwater runoff have reported reductions of 2 to 7 percent. Annual interception of rainfall by Sacramento's urban forest for the total urbanized area was only about 2 percent because of the winter rainfall pattern and sparsity of **evergreen** species (Xiao et al. 1998). However, average interception in canopied areas ranged from 6 to 13 percent (150 gal per tree), similar to values reported for rural forests. Broadleaf evergreens and **conifers** intercept more rainfall than deciduous species in areas where rainfall is highest in fall, winter, or spring (Xiao and McPherson 2002).

A recent study of the benefits of urban trees in Honolulu found that the 43,000 trees in the city's inventory intercepted 35 million gal of stormwater annually (Vargas et al. 2007a), valued at $350,000. The American Forests (2007) study of Palm Beach County, Florida, found that the 38-percent decline in the urban forest canopy between 2004 and 2006 owing to hurricanes meant that an additional 1 billion gal of stormwater had to be treated.

Urban forests can provide other hydrologic benefits, too. For example, when planted in conjunction with engineered soil around paved areas, trees can serve as mini stormwater reservoirs, capturing and filtering much more runoff than the trees alone (for more information, see our research summary *Engineered Soil, Trees and Stormwater Runoff* [CUFR 2007]). Tree plantations, nurseries, or landscapes can be irrigated with partially treated wastewater. Reused wastewater applied to urban forest lands can recharge aquifers, reduce stormwater-treatment loads, and create income through sales of nursery or wood products from the forests. Recycling urban wastewater into greenspace areas can be an economical means of treatment and disposal while at the same time providing other environmental benefits (USDA NRCS Agroforestry Center 2005).

Aesthetics and Other Benefits

Trees provide a host of aesthetic, social, economic, and health benefits that should be included in any benefit-cost analysis. One of the most frequently cited reasons that people plant trees is for beautification. Trees add color, texture, line, and form to the landscape, softening the hard geometry that dominates built environments. Research on the aesthetic quality of residential streets has shown that street trees are the single strongest positive influence on scenic quality (Schroeder and Cannon 1983).

In surveys, consumers have shown greater preference for commercial streetscapes with trees. In contrast to areas without trees, shoppers shop more often and longer in well-landscaped business districts. They are willing to pay more for parking and up to 11 percent more for goods and services (Wolf 1999).

Research in public housing complexes found that outdoor spaces with trees were used significantly more often than spaces without trees. By facilitating interactions among residents, trees can contribute to reduced levels of domestic violence, as well as foster safer and more sociable neighborhood environments (Sullivan and Kuo 1996).

Well-maintained trees increase the "curb appeal" of properties (fig. 8). Research documenting the increase in dollar value that can be attributed to trees is difficult to conduct and still in early stages, but some studies comparing sales prices of residential properties with different numbers of trees have suggested that people are willing to pay 3 to 7 percent more for properties with ample trees versus few or no trees. One of the most comprehensive studies of the influence of trees on home property values was based on actual sales prices in Athens, Georgia, and found that each large front-yard tree was associated with about a 1-percent increase in sales price (Anderson and Cordell 1988). A much greater value of 9 percent ($15,000) was determined in a U.S. Tax Court case for the loss of a large black oak on a

Figure 8—Trees beautify a neighborhood, increasing property values and creating a more sociable environment.

property valued at $164,500 (Neely 1988). Depending on average home sales prices, the value of this benefit can contribute significantly to cities' property tax revenues.

Scientific studies confirm that trees in cities provide social and psychological benefits. Humans derive substantial pleasure from trees, whether it is inspiration from their beauty, a spiritual connection, or a sense of meaning (Dwyer et al. 1992, Lewis 1996). After natural disasters, people often report a sense of loss if their community forest has been damaged (Hull 1992). Views of trees and nature from homes and offices provide restorative experiences that ease mental fatigue and help people to concentrate (Kaplan and Kaplan 1989). Desk workers with a view of nature report lower rates of sickness and greater satisfaction with their jobs compared to those having no visual connection to nature (Kaplan 1992). Trees provide important settings for recreation and relaxation in and near cities. The act of planting trees can have social value, as bonds between people and local groups often result.

A series of studies on human stress caused by general urban conditions show that views of nature reduce the stress response of both body and mind (Parsons et al. 1998), improving general well-being. Urban green also appears to have a positive effect on the human immune system. Hospitalized patients who have views of nature and spend time outdoors need less medication, sleep better, have a better outlook, and recover more quickly than patients without connections to nature (Ulrich 1985).

Skin cancer is a particular concern in the sunny, low-latitude Tropical region. By providing shade, trees reduce exposure to ultraviolet (UV) light, thereby lowering the risk of harmful effects from skin cancer and cataracts (Tretheway and Manthe 1999). In low-latitude regions like the Tropics, the ultraviolet protection factor (UPF) provided by trees increases from approximately 2 under a 30-percent canopy cover to as much as 30 under a 90-percent canopy cover (Grant et al. 2002). Because early exposure to UV radiation is a risk factor for later development of skin cancer, planting trees around playgrounds, schools, day care centers, and ball fields can be especially valuable in helping reduce the risk of later-life cancers.

Certain environmental benefits from trees are more difficult to quantify than those previously described, but can be just as important. Noise can reach unhealthy levels in cities. Trucks, trains, and planes can produce noise that exceeds 100 decibels (dB), twice the level at which noise becomes a health risk. Thick strips of vegetation in conjunction with landforms or solid barriers can reduce some highway noise and have a psychological effect (Cook 1978), but if vegetation is used as the only sound barrier, the amount necessary to achieve measurable reductions in noise (~200 ft for a 10-dB reduction) may be impractical (U.S. Department of Transportation 1995). Other studies have shown that the performance of noise barriers is increased when used in combination with vegetative screens (van Rentergehm et al. 2002).

Numerous types of wildlife inhabit cities and are generally highly valued by residents. For example, older parks, cemeteries, and botanical gardens often contain a rich assemblage of wildlife. Remnant woodlands and riparian habitats within cities can connect a city to its surrounding bioregion (fig. 9). Wetlands, greenways (linear parks), and other greenspace can provide habitats that conserve biodiversity (Platt et al. 1994).

Urban forestry can provide jobs for both skilled and unskilled labor. Public service programs and grassroots-led urban and community forestry programs provide horticultural training to volunteers across the United States. Also, urban and

Kelaine Vargas

Figure 9—Natural areas within cities can serve as refuges for wildlife and help connect city dwellers with their ecosystems.

community forestry provides educational opportunities for residents who want to learn about nature through firsthand experience (McPherson and Mathis 1999). Local nonprofit tree groups and municipal volunteer programs often provide educational material and hands-on training in the care of trees and work with area schools.

Tree shade on streets can help offset the cost of managing pavement by protecting it from weathering. The asphalt paving on streets contains stone aggregate in an oil binder. Tree shade lowers the street surface temperature and reduces heating and volatilization of the binder (McPherson and Muchnick 2005). As a result, the aggregate remains protected for a longer period by the oil binder. When unprotected, vehicles loosen the aggregate, and much like sandpaper, the loose aggregate grinds down the pavement. Because most weathering of asphalt-concrete pavement occurs during the first 5 to 10 years when new street tree plantings provide little shade, this benefit mainly applies when older streets are resurfaced (fig. 10).

Figure 10—Although shade trees can be expensive to maintain, their shade can reduce the costs of resurfacing streets (McPherson and Muchnick 2005), promote pedestrian travel, and improve air quality directly through pollutant uptake and indirectly through reduced emissions of volatile organic compounds from cars.

Costs

Planting and Maintaining Trees

The environmental, social, and economic benefits of urban and community forests come, of course, at a price. Although some general information is available on maintenance costs of street and park trees (see, e.g., Tschantz and Sacamano 1994), little has been published about the Tropical region. A recent study of the street trees of Honolulu estimated their annual maintenance costs at $30 per tree (Vargas et al. 2007a).

Annual expenditures for tree management on private property have also not been well documented. Costs differ considerably, ranging from some commercial or residential properties that receive regular professional landscape service to others that are virtually "wild" and without maintenance. An analysis of data for Sacramento suggested that households typically spent about $5 to $10 annually per tree for pruning and pest and disease control (Summit and McPherson 1998). Expenditures are usually greatest for pruning, planting, and removal.

Conflicts With Urban Infrastructure

Like other cities across the United States, communities in the Tropical region are spending millions of dollars each year to manage conflicts between trees and power lines, sidewalks, sewers, and other elements of the urban infrastructure. A recent study showed that one Tropical community is spending about $10 per tree annually on sidewalk, curb, and gutter repair costs (Vargas et al. 2007a). This amount is close to the value of $11.22 per tree reported for 18 California cities (McPherson 2000).

In some cities, decreasing budgets are increasing the sidewalk-repair backlog and forcing cities to shift the costs of sidewalk repair to residents. This shift has significant impacts on residents in older areas, where large trees have outgrown small sites and infrastructure has deteriorated. It should be noted that trees are not always solely responsible for these problems. In older areas, in particular, sidewalks and curbs may have reached the end of their 20- to 25-year service life, or may have been poorly constructed in the first place (Sydnor et al. 2000).

Efforts to control the costs of these conflicts are having alarming effects on urban forests (Bernhardt and Swiecki 1993, Thompson and Ahern 2000):

- Cities are downsizing their urban forests by planting smaller trees. Although small trees are appropriate under power lines and in small planting sites, they are less effective than large trees at providing shade, absorbing air pollutants, and intercepting rainfall.
- Thousands of healthy urban trees are lost each year and their benefits forgone because of sidewalk damage, the second most common reason that street and park trees were removed.
- Most cities surveyed were removing more trees than they were planting. Residents forced to pay for sidewalk repairs may not want replacement trees.

Cost-effective strategies to retain benefits from large street trees while reducing costs associated with infrastructure conflicts are described in *Reducing Infrastructure Damage by Tree Roots* (Costello and Jones 2003). Matching the growth characteristics of trees to the conditions at the planting site is one important strategy.

Tree roots can also damage old sewer lines that are cracked or otherwise susceptible to invasion. Sewer repair companies estimate that sewer damage is minor until trees and sewers are over 30 years old, and roots from trees in yards are usually more of a problem than roots from trees in planter strips along streets. The latter assertion may be because the sewers are closer to the root zone as they enter houses than at the street. Repair costs typically range from $100 for sewer rodding (inserting a cleaning implement to temporarily remove roots) to $1,000 or more for sewer excavation and replacement.

Most communities sweep their streets regularly to reduce surface-runoff pollution entering local waterways. Street trees drop leaves, flowers, fruit, and branches year round that constitute a significant portion of debris collected from city streets. When leaves fall and rains begin, **tree litter** can clog sewers, dry wells, and other elements of flood-control systems. Costs include additional labor needed to remove leaves, and property damage caused by localized flooding. In Tropical communities, hurricanes contribute to higher than average cleanup costs.

The cost of addressing conflicts between trees and power lines is reflected in electric rates. Large trees under power lines require more frequent pruning than better-suited trees, which can make them appear less attractive (fig. 11). Frequent crown reduction reduces the benefits these trees could otherwise provide. Moreover, increased costs for pruning are passed on to customers.

Shauna Cozad

Figure 11—Large trees planted under power lines can require extensive pruning, which increases tree care costs and reduces the benefits of those trees, including their appearance.

Chapter 3. Benefits and Costs of Community Forests in Tropical Communities

This chapter presents estimated benefits and costs for trees planted in typical residential yards and public sites in Tropical communities. Because benefits and costs differ with tree size, we report results for representative small, medium, and large trees.

Estimates are initial approximations as some benefits and costs are intangible or difficult to quantify (e.g., impacts on psychological health, crime, and violence). Limited knowledge about physical processes at work and their interactions makes estimates imprecise (e.g., fate of air pollutants trapped by trees and then washed to the ground by rainfall). Tree growth and mortality rates are highly variable throughout the region. Benefits and costs also differ, depending on differences in climate, pollutant concentrations, maintenance practices, and other factors. Given the Tropical region's diverse landscape, with different climates, soils, and types of community forestry programs, the approach used here provides first-order approximations. It is a general accounting that can be easily adapted and adjusted for local planting projects. It provides a basis for decisions that set priorities and influence management direction (Maco and McPherson 2003).

Overview of Procedures

Approach

In this study, annual benefits and costs are estimated over a 40-year planning horizon for newly planted trees in three residential yard locations (east, south, and west of the residence) and a public streetside or park location (app. 2). Henceforth, we refer to trees in these hypothetical locations as "yard" trees and "public" trees, respectively. Trees planted in locations other than yards and along streets—for example, in commercial, industrial, or institutional areas—provide benefits as well; however, the effects of trees on energy conservation and property values for nonresidential buildings have not been studied indepth and cannot be adequately modeled. It can be conservatively assumed that the benefits of trees in these locations lie somewhere between the yard and public values provided here.

Prices are assigned to each cost (e.g., planting, pruning, removal, infrastructure repair) and benefit (e.g., cooling energy savings, air pollutant mitigation, stormwater runoff reduction, and aesthetic and other benefits measured as increases in property value) through direct estimation and implied valuation of benefits as environmental externalities. This approach makes it possible to estimate the net benefits of plantings in "typical" locations by using "typical" tree species. More information on data collection, modeling procedures, pricing of costs and benefits, and assumptions can be found in appendix 3.

To account for differences in the mature size and growth of different tree species, we report results for a small (silver buttonwood), medium (rainbow shower tree), and large (monkeypod) tree (figs. 12 to 14) (see "Common and Scientific Names" section). The selection of these species is based on data availability and representative growth and is not necessarily intended to endorse their use in large numbers.

There has been controversy in recent years about the magnitude of the environmental and other benefits of palm trees. Some argue that they have little value and should be avoided in favor of shade trees. Others point to their aesthetic value and their role in creating a "sense of place" in arguing for their use. It is clear that palm trees, especially larger ones, provide shade for cooling, sequester carbon, remove air pollutants from the air, and trap stormwater. At the same time, they can be very expensive to plant and maintain. Coconut palms, for example, must have their coconuts removed each year to protect pedestrians and vehicles. The relative cost-effectiveness of palm trees is unresolved and that research question is beyond the scope of our work here. Our research has long had "true" trees as its emphasis, and they are therefore the focus of this guide.

Tree dimensions are derived from growth curves developed from street trees in Honolulu, Hawaii (Vargas et al. 2007a) (fig. 15). Frequency and costs of tree management are estimated based on data from municipal foresters in Honolulu, Hawaii, and Naples and Lantana, Florida. In addition, commercial arborists from Ewa Beach, Honolulu, Kaneohe, and Waimanalo, Hawaii, and Hollywood, Florida, provided information on tree management costs on residential properties.

Benefits are calculated with numerical models and data both from the region (e.g., pollutant emission factors for avoided emissions from energy savings) and from local sources (e.g., Honolulu climate data for energy effects). An average of regional electricity prices is used in this study to quantify energy savings. **Damage costs** and **control costs** are used to estimate **willingness to pay.** For example, the value of stormwater runoff reduction owing to rainfall interception by trees is estimated by using marginal control costs. If a community or developer is willing to pay an average of $0.01 per gal of treated and controlled runoff to meet minimum standards, then the stormwater runoff mitigation value of a tree that intercepts 1,000 gal of rainfall, eliminating the need for control, should be $10.

Figure 12—The silver buttonwood represents small trees in this guide.

Figure 13—The rainbow shower tree represents medium trees in this guide.

Figure 14—The monkeypod represents large trees in this guide.

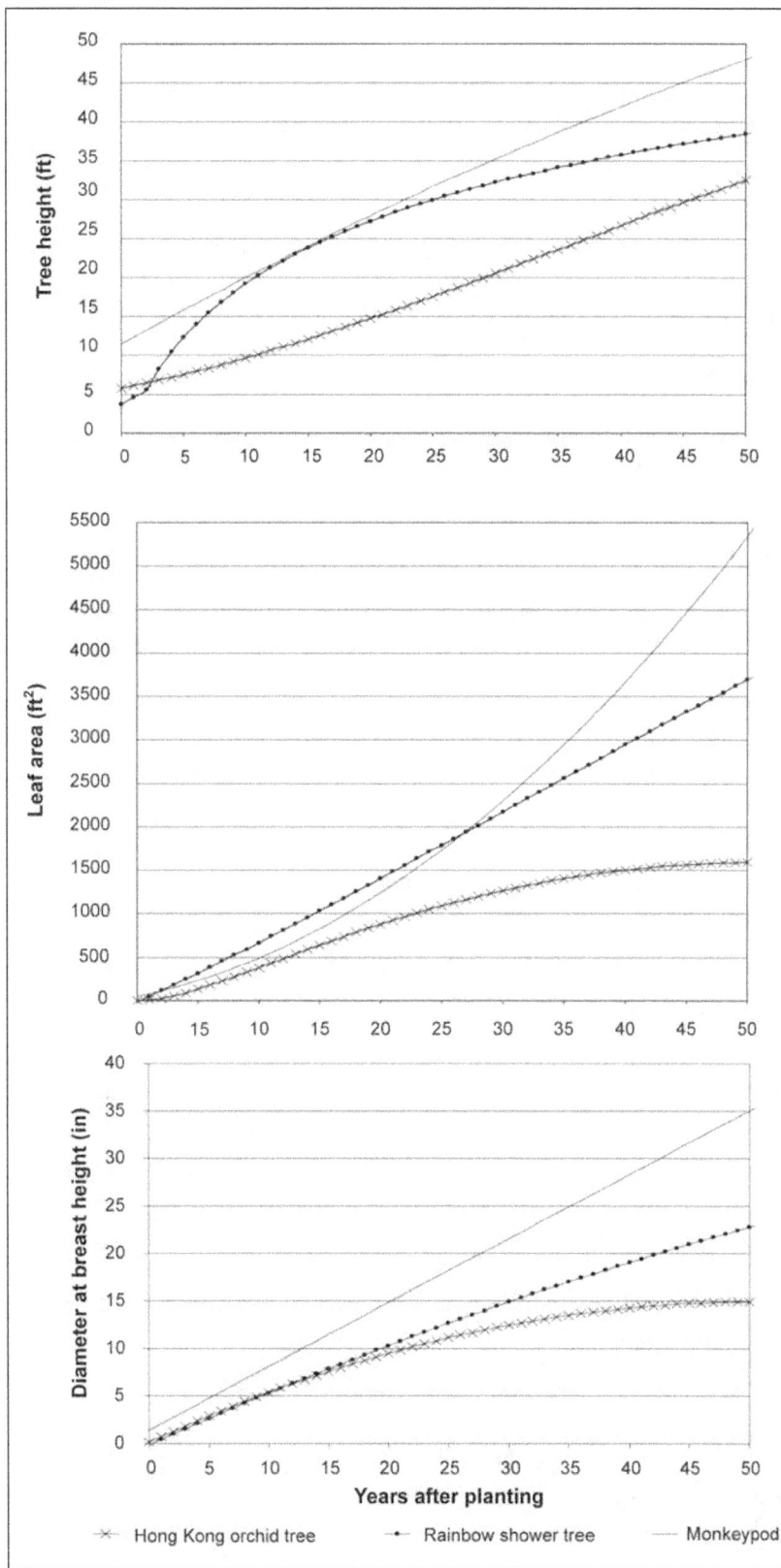

Figure 15—Tree growth curves are based on data collected from park trees in Honolulu, Hawaii. Data for representative small, medium, and large trees are for the silver buttonwood, rainbow shower tree, and monkeypod, respectively. Differences in leaf surface area among species are most important for this analysis because functional benefits such as summer shade, rainfall interception, and pollutant uptake are related to leaf area.

27

Reporting Results

Results are reported in terms of annual value per tree planted. To make these calculations realistic, however, mortality rates are included. Based on our survey of regional municipal foresters and commercial arborists, this analysis assumes that 19 percent of the planted trees will die over the 40-year period. Annual mortality rates are 1.0 percent per year for the first 5 years and 0.4 percent per year for the remainder of the 40-year period. This accounting approach "grows" trees in different locations and uses computer simulation to calculate the annual flow of benefits and costs as trees mature and die (McPherson 1992). In appendix 2, results are reported at 5-year intervals for 40 years.

Findings of This Study

Average Annual Net Benefits

Average annual net benefits (benefits minus costs) per tree over a 40-year period increase with mature tree size (for detailed results see app. 2):

- $9 to $30 for a small tree
- $43 to $79 for a medium tree
- $70 to $92 for a large tree

Our findings demonstrate that average annual net benefits from large trees like the monkeypod are substantially greater than those from small trees like the silver buttonwood. Public trees represent the lower end of the range: $9, $43, and $70 for small, medium, and large trees, respectively. Yard trees opposite the west-facing wall of a house represent the higher end of the range: $30, $79, and $92, for small, medium, and large trees, respectively.

At year 40, the large yard tree opposite a west wall produces a net annual benefit of $168. In the same location, 40 years after planting, the silver buttonwood and rainbow shower tree produce annual net benefits of $30 and $79. Forty years after planting at a typical public site, the small, medium, and large trees provide annual net benefits of $9, $43, and $70, respectively.

Net benefits for a yard tree opposite a west house wall and a public tree also increase with size when summed over the entire 40-year period:

- $1,330 (yard) and $340 (public) for a small tree
- $3,535 (yard) and $1,905 (public) for a medium tree
- $4,115 (yard) and $3,060 (public) for a large tree

Even just 20 years after planting, average annual benefits for all trees exceed costs of tree planting and management (tables 1 and 2). For a large monkeypod in a yard 20 years after planting, the total value of environmental benefits alone ($47) is more than five times the total annual cost ($9). Environmental benefits at year 20 are $15 and $50 for the silver buttonwood tree and rainbow shower tree, and tree care costs are $2 and $8 for each species, respectively. Adding the value of aesthetics and other benefits to the environmental benefits results in even greater net benefits.

Net benefits for public trees at 20 years ($11, $41, and $57 for small, medium, and large trees; table 2) are less than yard trees ($28, $83, and $86) for two main reasons: public tree care costs are greater because public trees generally receive more intensive care than private trees; and energy benefits are lower for public trees than for yard trees because public trees are assumed to provide general climate effects, but not to shade buildings directly.

Average Annual Costs

Averaged over 40 years, the costs for yard and public trees, respectively, are as follows:

- $12 and $24 for a small tree
- $14 and $29 for a medium tree
- $16 and $35 for a large tree

Annualized over the 40-year period, tree planting is the single greatest cost for yard trees, averaging $7.50 per tree per year (see app. 2). Based on our survey, we assume in this study that a 25-gal yard tree is planted at a cost of $300; the price includes the tree, labor, and any necessary watering during the establishment period. The cost for planting a 25-gal public tree is $120 ($3 per year). For public trees, pruning ($10 to $17 per tree per year) and managing conflicts with infrastructure ($5 to $8 per tree per year) are the greatest costs. Annual pruning costs are also significant for yard trees, ranging from $2 to $5. Removal and disposal costs, annualized over 40 years, average $1 to $2 per tree. At $2 to $3 per tree per year, administrative costs are significant for public trees, as are cleanup costs ($2 to $3 per tree) in this hurricane-prone region.

Table 3 shows annual management costs 20 years after planting for yard trees to the west of a house and for public trees. Annual costs for yard trees range from $2 to $9, and public tree care costs are $16 to $34. In general, public trees are more expensive to maintain than yard trees because of their prominence, the greater need for public safety, and conflicts with infrastructure.

Table 1—Estimated annual benefits and costs for a private tree (residential yard) opposite the west-facing wall 20 years after planting

Benefit category	Silver buttonwood small tree 15 ft tall 17-ft spread LSA = 341 ft²		Rainbow shower tree medium tree 27 ft tall 27-ft spread LSA = 1,419 ft²		Monkeypod large tree 28 ft tall 36-ft spread LSA = 1,245 ft²	
	Resource units	Total value	Resource units	Total value	Resource units	Total value
		Dollars		*Dollars*		*Dollars*
Electricity savings ($0.122/kWh)	73 kWh	8.95	273 kWh	33.26	207 kWh	25.23
Carbon dioxide ($0.003/lb)	179 lb	0.60	525 lb	1.75	431 lb	1.44
Ozone ($1.47/lb)	0.12 lb	0.17	0.27 lb	0.40	0.46 lb	0.68
Nitrogen dioxide ($1.47/lb)	0.27 lb	0.40	0.97 lb	1.43	0.96 lb	1.42
Sulfur dioxide ($1.52/lb)	0.24 lb	0.36	0.85 lb	1.30	0.84 lb	1.28
Small particulate matter ($1.34/lb)	0.18 lb	0.24	0.44 lb	0.59	0.48 lb	0.64
Volatile organic compounds ($0.60/lb)	0.04 lb	0.03	0.16 lb	0.10	0.16 lb	0.10
Biogenic volatile organic compounds ($0.60/lb)	-0.01 lb	-0.01	-0.01 lb	-0.01	-0.01 lb	-0.01
Rainfall interception ($0.01/gal)	425 gal	4.25	1,100 gal	11.00	1,585 gal	15.85
Environmental subtotal		15.00		49.83		46.63
Aesthetics and other benefits		14.98		40.71		47.89
Total benefits		29.97		90.54		94.52
Total costs		2.38		7.60		8.62
Net benefits		27.59		82.94		85.90

LSA = leaf surface area.

Table 2—Estimated annual benefits and costs for a public tree (street/park) 20 years after planting

Benefit category	Silver buttonwood small tree 15 ft tall 17-ft spread LSA = 341 ft²		Rainbow shower tree medium tree 27 ft tall 27-ft spread LSA = 1,419 ft²		Monkeypod large tree 28 ft tall 36-ft spread LSA = 1,245 ft²	
	Resource units	Total value	Resource units	Total value	Resource units	Total value
		Dollars		*Dollars*		*Dollars*
Electricity savings ($0.122/kWh)	28 kWh	3.41	71 kWh	8.69	127 kWh	15.46
Carbon dioxide ($0.003/lb)	100.44 lb	0.34	176.30 lb	0.59	292.07 lb	0.98
Ozone ($1.47/lb)	0.12 lb	0.17	0.27 lb	0.40	0.46 lb	0.68
Nitrogen dioxide ($1.47/lb)	0.27 lb	0.40	0.97 lb	1.43	0.96 lb	1.42
Sulfur dioxide ($1.52/lb)	0.24 lb	0.36	0.85 lb	1.30	0.84 lb	1.28
Small particulate matter ($1.34/lb)	0.18 lb	0.24	0.44 lb	0.59	0.48 lb	0.64
Volatile organic compounds ($0.60/lb)	0.04 lb	0.03	0.16 lb	0.10	0.16 lb	0.10
Biogenic volatile organic compounds ($0.60/lb)	0 lb	-0.01	0 lb	-0.01	0 lb	-0.01
Rainfall interception ($0.01/gal)	425 gal	4.25	1,100 gal	11.00	1,585 gal	15.85
Environmental subtotal		9.20		24.09		36.39
Aesthetics and other benefits		16.97		46.11		54.25
Total benefits		26.16		70.20		90.64
Total costs		15.63		29.20		33.68
Net benefits		10.53		41.01		56.96

LSA = leaf surface area.

Table 3—Estimated annual costs 20 years after planting for a private tree opposite the west-facing wall and a public tree

Costs	Silver buttonwood small tree 15 ft tall 17-ft spread LSA = 341 ft^2		Rainbow shower tree medium tree 27 ft tall 27-ft spread LSA = 1,419 ft^2		Monkeypod large tree 28 ft tall 36-ft spread LSA = 1,245 ft^2	
	Private: west	Public tree	Private: west	Public tree	Private: west	Public tree
	Dollars per tree per year					
Tree and planting	0	0	0	0	0	0
Pruning	0.14	5.85	5.26	19.00	5.26	19.00
Remove and dispose	1.40	0.89	1.46	0.93	2.10	1.34
Pest and disease	0.00	0.26	0.00	0.27	0.00	0.39
Infrastructure	0.63	5.05	0.66	5.27	0.95	7.58
Irrigation	0	0	0	0	0	0
Cleanup	0.21	1.68	0.22	1.76	0.32	2.53
Liability and legal	0	0	0	0	0	0
Administration and other	0	1.89	0	1.97	0	2.84
Total costs	2.38	15.63	7.60	29.20	8.62	33.68
Total benefits	29.97	26.16	90.54	70.20	94.52	90.64
Total net benefits	27.59	10.53	82.94	41.01	85.90	56.96

LSA = leaf surface area.

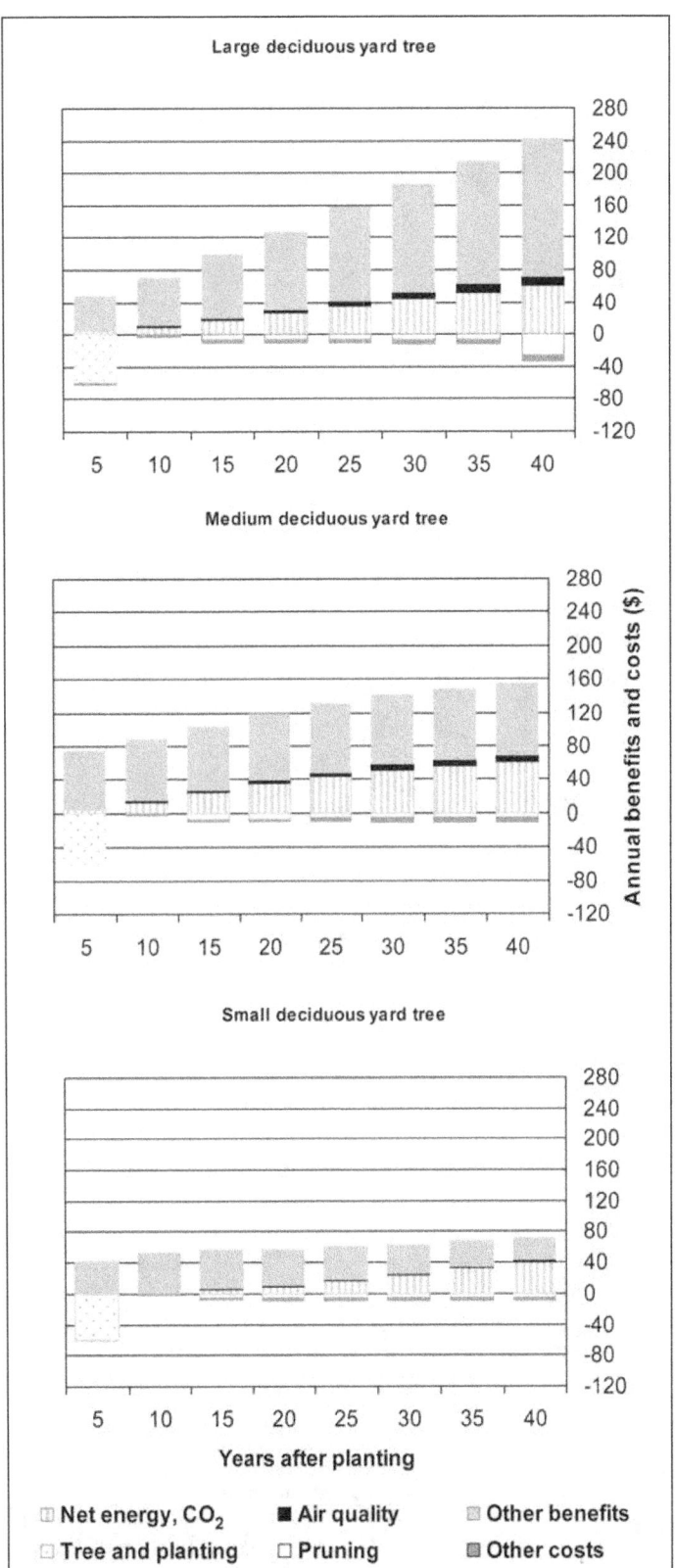

Figure 16—Estimated annual benefits and costs (at 5-year intervals) for a small (silver buttonwood), medium (rainbow shower tree), and large (monkeypod) tree located west of a residence. Costs are greatest during the initial establishment period, whereas benefits increase with tree size.

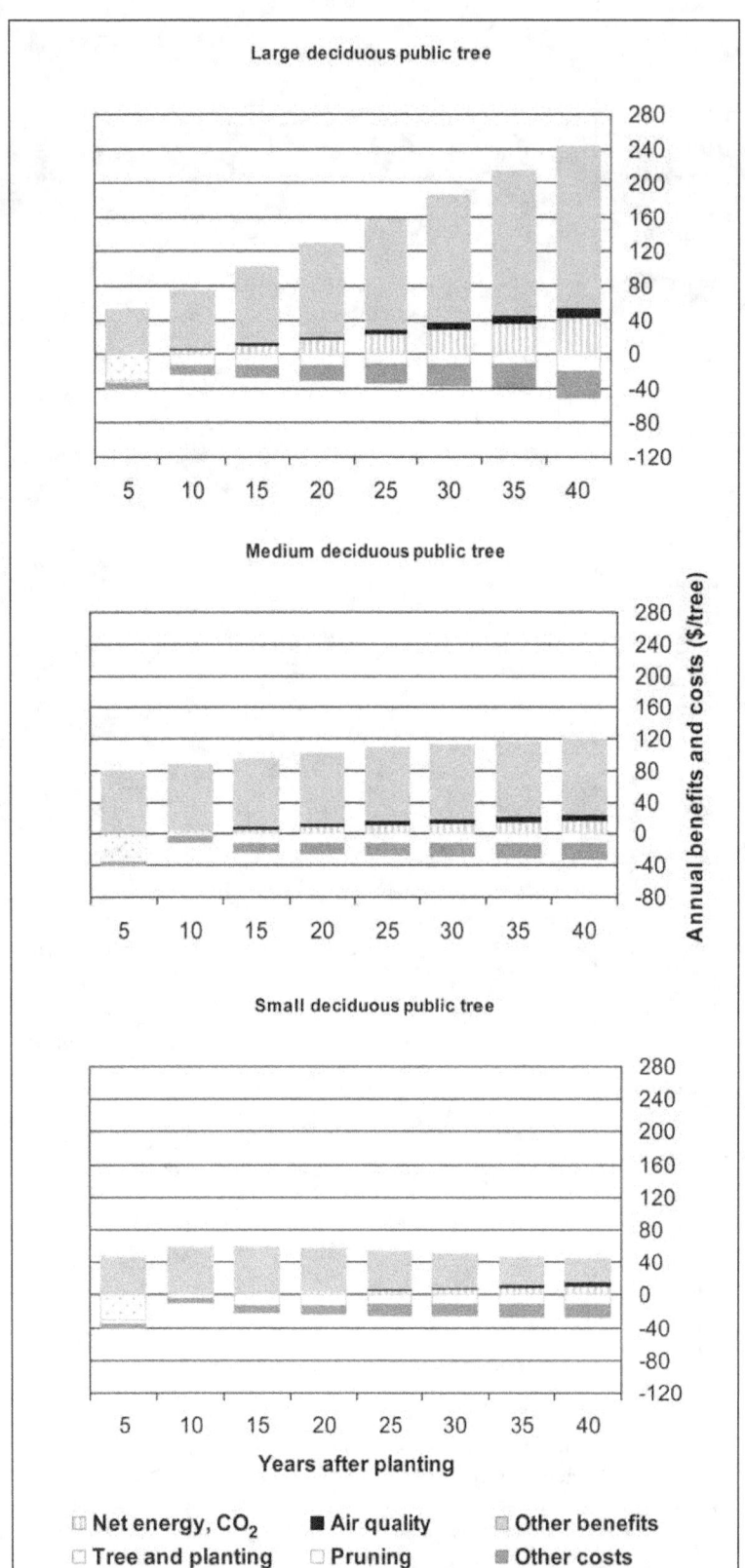

Figure 17—Estimated annual benefits and costs (at 5-year intervals) for public small (silver buttonwood), medium (rainbow shower tree), and large (monkeypod) tree.

Average Annual Benefits

Average annual benefits, including stormwater reduction, aesthetic value, air quality improvement and carbon dioxide (CO_2) sequestration increase with mature tree size (figs. 16 and 17; for detailed results see app. 2):

- $33 to $42 for a small tree
- $72 to $92 for a medium tree
- $105 to $108 for a large tree

Energy savings—

Energy benefits are the most significant environmental benefit and tend to increase with mature tree size. For example, average annual energy benefits over the 40-year period are $15 for the small silver buttonwood tree opposite a west-facing wall and $29 for the larger monkeypod. For species of all sizes, energy savings increase as trees mature and their leaf surface area increases (figs. 16 and 17).

As might be expected in the Tropical region, cooling savings are substantial. Trees planted on the west side of buildings have the greatest energy benefits because the effect of shade on cooling costs is maximized by blocking the sun during the warmest time of the day. A yard tree located south of a home produced the least total benefit because, at the lower latitudes of the Tropical region where the Sun remains mostly overhead throughout the year, little sunlight strikes a building on the south side. Trees located east of a building provided intermediate benefits. Total energy benefits also reflect species-related traits such as size, form, branch pattern, and density, as well as time in leaf.

Average annual total energy benefits for public trees were less than for yard trees and ranged from $4 for the silver buttonwood tree to $20 for the large monkeypod.

Stormwater runoff reduction—

By intercepting rain before it reaches the stormwater treatment system, trees can have a valuable effect on reducing runoff. The monkeypod intercepts 2,108 gal per year on average over a 40-year period with an implied value of $21. The silver buttonwood and rainbow shower tree intercept 605 and 1,237 gal per year on average, with values of $6 and $12, respectively. Forty years after planting, average stormwater runoff reductions equal 1,387, 2,270, and 4,529, respectively, for the small, medium, and large trees. The hydrology modeling was based on rainfall data for the Honolulu International Airport, which has the lowest levels of rainfall on the island of Oahu (17 in per year); other, wetter parts of the Tropical region can expect much higher benefits.

As the cities of the Tropical region continue to grow, the amount of impervious surface will continue to increase. The role that trees, in combination with other strategies such as rain gardens and structural soils, can play in reducing stormwater runoff is substantial.

Air quality improvement—

Air quality benefits are defined as the sum of pollutant uptake by trees and avoided powerplant emissions from energy savings minus biogenic volatile organic compounds released by trees. Average annual air quality benefits over the 40-year period were approximately $2 to $5 per tree. These relatively low air quality benefits reflect the clean air of most cities in the Tropical region. Contrast these results with the air quality benefits of a large tree in the Northeast ($13; McPherson et al. 2007), Midwest ($7.65; McPherson et al. 2006c), and southern California ($28.38; McPherson et al. 2000).

The ability of trees to intercept nitrogen dioxide (NO_2) from the air is the most highly valued. Over 40 years, the monkeypod, for example, is estimated to reduce an average of 1.18 lb of NO_2 from the air annually, valued at $1.74. Average annual reductions in sulfur dioxide, small particulate matter, ozone, and volatile organic compounds for the large tree are valued at $1.58, $0.98, $0.88, and $0.12, respectively.

Forty years after planting, the average annual monetary values of air quality improvement (avoided + uptake of pollutants) for the silver buttonwood, rainbow shower tree, and monkeypod are $4.40, $7.31, and $11.08, respectively.

Carbon dioxide reduction—

Net atmospheric CO_2 reductions accrue for all tree types. Average annual net reductions range from a high of 508 lbs ($1.70) for a large tree on the west side of a house to a low of 174 lbs ($1.08) for a small public tree. Deciduous trees opposite west-facing house walls generally produce the greatest CO_2 reduction from reduced powerplant emissions associated with energy savings. The values for the silver buttonwood tree are lowest for CO_2 reduction reflecting this small tree's minor effect on energy savings and sequestration.

Forty years after planting, net CO_2 benefits for a yard tree opposite a west wall are 841, 900, and 982 lbs, respectively, for the small, medium, and large trees. Releases of CO_2 associated with tree care activities account for less than 1 percent of net CO_2 sequestration.

Aesthetic and other benefits—

Aesthetic and other benefits reflected in property values account for the greatest portion of total benefits. As trees grow and become more visible, they can increase a property's sales price. Annual values averaged over 40 years associated with these aesthetic and other benefits for yard trees are $18, $40, and $51 for the small, medium, and large trees, respectively. The values for public trees are $20, $45, and $57, respectively. The values for yard trees are slightly less than for public trees because off-street trees contribute less to a property's curb appeal than more prominent street trees. Because our estimates are based on median home sale prices, the effects of trees on property values and aesthetics will differ depending on local economies.

The green infrastructure is a significant component of communities in the Tropical region. A Bahamian town is shown here.

Chapter 4. Estimating Benefits and Costs for Tree Planting Projects in Your Community

This chapter shows two ways that benefit-cost information presented in this guide can be used. The first hypothetical example demonstrates how to adjust values from the guide for local conditions when the goal is to estimate benefits and costs for a proposed tree planting project. The second example explains how to compare net benefits derived from planting different types of trees. The last section discusses actions communities can take to increase the cost-effectiveness of their tree programs.

Applying Benefit-Cost Data

Punakea Valley Example

The hypothetical city of Punakea Valley is located in the Tropical region and has a population of 74,000. Most of its street trees were planted decades ago, with West Indian mahogany and coral trees (see "Common and Scientific Names" section) as the dominant species. Currently, the tree canopy cover is sparse because a recent hurricane and a pest destroyed many of the trees and they have not been replaced. Many of the remaining street trees are in declining health. The city hired an urban forester 2 years ago, and an active citizens' group, the Green Team, has formed (fig. 18).

Initial discussions among the Green Team, local utilities, the urban forester, and other partners led to a proposed urban forestry program. The program intends to plant 1,000 trees in Punakea Valley over a 5-year period. Trained volunteers will plant 2-in-diameter trees in the following proportions: 50 percent large-maturing trees, 35 percent medium-maturing trees, and 15 percent small-maturing trees. One hundred trees will be planted in parks, and the remaining 900 trees will be planted along Main Street and other downtown streets. Mortality rates for earlier planting projects have been high, so the Green Team and the urban forester will concentrate their planting efforts in areas that are likely to be most successful, including planting spaces with sufficient soil capacity for trees to grow and as little conflict with infrastructure as possible, and that maximize environmental benefits. They expect to find a number of good suggestions for planting in chapter 5 of this guide.

The Punakea Valley City Council has agreed to maintain the current funding level for management of existing trees. Also, they will advocate formation of a municipal tree district to raise funds for the proposed tree-planting project. A

Figure 18—The (hypothetical) Green Team is motivated to re-green their community by planting 1,000 trees in 5 years.

municipal tree district is similar in concept to a landscape assessment district, which receives revenues based on formulas that account for the services different customers receive. For example, the proximity of customers to greenspace in a landscape assessment district may determine how much they pay for upkeep. A municipal tree district might receive funding from air quality districts, stormwater management agencies, electric utilities, businesses, and residents in proportion to the value of future benefits these groups will receive from trees in terms of air quality, hydrology, energy, carbon dioxide (CO_2), and property value. The formation of such a district would require voter approval of a special assessment that charges recipients for tree planting and maintenance costs in proportion to the benefits they receive from the new trees. The council needs to know the amount of funding required for tree planting and maintenance, as well as how the benefits will be distributed over the 40-year life of the project.

As a first step, the Punakea Valley city forester and Green Team decided to use the tables in appendix 2 to quantify total cumulative benefits and costs over 40 years for the proposed planting of 1,000 public trees—500 large, 350 medium, and 150 small deciduous.

Before setting up a spreadsheet to calculate benefits and costs, the team considered which aspects of Punakea Valley's urban and community forestry project differ from the regional values used in this guide (the methods for calculating the values in app. 2 are described in app. 3):

1. The prices of electricity and natural gas in Punakea Valley are $0.26 per kWh, not $0.122 per kWh as used in this guide. It is assumed that the buildings that will be shaded by the new street trees have air conditioning.

2. The Green Team projected future annual costs for monitoring tree health and implementing their stewardship program. Administration costs are estimated to average $2,500 annually for the life of the trees or $2.50 per tree each year. This guide assumed an average annual administration cost of about $2.00 per tree. Thus, an adjustment is necessary.

3. Planting will cost $200 per tree. The guide assumes planting costs of $120 per tree. The costs will be slightly higher for Punakea Valley because they have decided to plant larger trees.

To calculate the dollar value of total benefits and costs for the 40-year period, the forester created a spreadsheet table (table 4). Each benefit and cost category is listed in the first column. Prices, adjusted where necessary for Punakea Valley, are entered into the second column. The third column contains the **resource units** (RUs) per tree per year associated with the benefit or the cost per tree per year, which can be found in appendix 2. For aesthetic and other benefits, the dollar values for public trees are placed in the RU columns. The fourth column lists the 40-year total values, obtained by multiplying the RU values by tree numbers, prices, and 40 years.

To adjust for higher electricity prices, the forester multiplied electricity saved for a large public tree in the RU column (162 kWh) by the Punakea Valley price for electricity ($0.26/kWh). This value ($42.07 per tree per year) was multiplied by the number of trees planted and 40 years ($42.07 × 500 trees × 40 years = $842,400) to obtain cumulative air-conditioning energy savings for the large public trees (table 4). The process was carried out for all benefits and all tree types.

To adjust cost figures, the city forester changed the planting cost from $120 assumed in the Guide to $200 (table 4). This planting cost was annualized by dividing the cost per tree by 40 years ($200/40 = $5 per tree per year). Total planting costs were calculated by multiplying this value by 500 large trees and 40 years ($200,000).

The administration, inspection, and outreach costs are expected to average $2.50 per tree per year. Consequently, the total administration cost for large trees is

Table 4—Spreadsheet calculations of benefits and costs for the Punakea Valley planting project (1,000 trees) over 40 years

	Adjusted price	150 small trees		350 medium trees		500 large trees		1,000 total trees		
		Resource units	Total value	Resource units	Total value	Resource units	Total value	Total value	Annual value per tree	Percentage of benefits
	Dollars	RU/tree/yr	Dollars	RU/tree/yr	Dollars	RU/tree/yr	Dollars	Dollars	$/tree/yr	
Benefits:										
Electricity (kWh)	0.26	37	57,720	78	283,920	162	842,400	1,184,040	29.60	30
Net carbon dioxide (lb)	0.0033	174	3,487	188	8,791	370	24,716	36,994	0.92	1
Ozone (lb)	1.47	0.16	1,414	0.30	6,185	0.60	17,671	25,269	0.63	—
Nitrogen dioxide (lb)	1.47	0.45	3,976	1.03	21,234	1.18	34,753	59,963	1.50	—
Sulfur dioxide (lb)	1.52	0.39	3,567	0.90	19,206	1.03	31,400	54,172	1.35	—
Small particulate matter (lb)	1.34	0.25	2,008	0.51	9,558	0.73	19,545	31,111	0.78	4.5
Volatile organic compounds (lb)	0.60	0.08	289	0.17	1,434	0.19	2,289	4,012	0.10	—
Biogenic volatile organic compounds (lb)	0.60	-0.01	-36	-0.02	-169	-0.02	-241	-446	-0.01	—
Hydrology (gal)	0.01	605	36,300	1,237	173,180	2,108	421,600	631,080	15.78	16
Aesthetics and other		5.00	119,880	45.01	630,140	57.27	1,145,400	1,895,420	47.39	48.5
Total benefits			228,605		1,153,479		2,539,533	3,921,615	98.04	100.00
Costs:										
Tree and planting ($)		9.92	30,000	5.00	70,000	5.00	100,000	200,000	5.00	15
Pruning ($)		1.02	59,534	15.86	221,975	16.91	338,193	619,702	15.49	47
Remove and dispose ($)		5.53	6,116	0.93	13,052	1.40	28,056	47,225	1.18	4
Infrastructure repair ($)		1.84	33,162	5.07	71,039	7.51	150,267	254,468	6.36	19.5
Cleanup ($)		2.50	11,054	1.69	23,680	2.50	50,089	84,823	2.12	6.5
Administration and other ($)		2.50	15,000	2.50	35,000	2.50	50,000	100,000	2.50	8
Total costs			154,866		434,745		716,606	1,306,217	32.66	100.00
Net benefit			73,739		718,734		1,822,927	2,615,397	65.38	
Benefit/cost ratio			1.47		2.65		3.54	3.00		

RU = resource units.

2.50×500 large trees $\times 40$ years ($50,000). The same procedure was followed to calculate costs for the medium and small trees.

All costs and all benefits were summed. Annual benefits over 40 years for the whole planting total $3.9 million ($98.04 per tree per year), and annual costs total about $1.3 million ($32.66 per tree per year). Subtracting total costs from total benefits yields net benefits over the 40-year period:

- $73,739 or $12.29 per tree per year for small trees
- $718,734 or $51.34 per tree per year for medium trees
- $1.8 million or $91.15 per tree per year for large trees
- $2.6 million or $65.31 per tree per year for the project

Dividing total benefits by total costs yielded benefit-cost ratios (BCRs) that ranged from 1.47 for small trees to 2.65 and 3.54 for medium and large trees. The BCR for the entire planting is 3.00, indicating that $3.00 of value will be returned for every $1 invested.

This analysis assumes 19 percent of the planted trees die and does not account for the time value of money from a capital investment perspective. Use the municipal discount rate to compare this investment in tree planting and management with alternative municipal investments.

The city forester and Green Team now know that the project will cost about $1.3 million, and the average annual cost will be $32,655 ($1.3 million/40 years), although a higher proportion of funds will be needed initially for planting. The fifth and last step is to identify the distribution of functional benefits that the trees will provide. The last column in table 4 shows the distribution of benefits as a percentage of the total:

- Energy savings = 30 percent
- CO_2 reduction = 1 percent
- Stormwater runoff reduction = 16 percent
- Air quality improvement = 4.5 percent
- Aesthetics/property value increase = 48.5 percent

With this information the planning team can determine how to distribute the costs for tree planting and maintenance based on who benefits from the services the trees will provide. For example, assuming the goal is to generate enough annual revenue to cover the total costs of managing the trees ($1.3 million), fees could be distributed in the following manner:

- $390,000 from electric and natural gas utilities for peak energy savings (30 percent). (Utility companies invest in planting trees because it is more cost effective to reduce peak energy demand than to meet peak needs through added infrastructure.)

- $13,000 from local industry for atmospheric CO_2 reductions (1 percent).
- $208,000 from the stormwater management district for water quality improvement associated with reduced runoff (16 percent).
- $58,500 from air quality management district for net reduction in air pollutants (4.5 percent).
- $630,500 from property owners for increased property values (64 percent).

Whether funds are sought from partners, the general fund, or other sources, this information can assist managers in developing policy, setting priorities, and making decisions. The Center for Urban Forest Research has developed a computer program called STRATUM that simplifies these calculations for analysis of existing street tree populations (Maco and McPherson 2003; for more information, see www.itreetools.org [USDA FS 2006a]).

City of Mangrovia Example

Ten years ago, as a municipal cost-cutting measure, the hypothetical city of Mangrovia stopped planting street trees in areas of new development. Instead, developers were required to plant front yard trees, thereby reducing costs to the city. The community forester and concerned citizens came to notice that instead of the large, stately trees the city had once planted, developers were planting small flowering trees, which were more aesthetically pleasing in early years, but would never achieve the stature—or the benefits—of larger shade trees. To evaluate the consequences of these changes, the community forester and citizens decided to compare the benefits of planting small, medium, and large trees for a hypothetical street-tree planting project in a new neighborhood in Mangrovia.

As a first step, the city forester and concerned citizens decided to quantify the total cumulative benefits and costs over 40 years for three potential street tree planting scanarios in Mangrovia. The scenarios compare plantings of 500 small trees, 500 medium trees, and 500 large trees. Data in appendix 2 are used for the calculations; however, three aspects of Mangrovia's urban and community forestry program are different from those assumed in this tree guide:

1. The price of electricity is $0.062/kWh, not $0.122/kWh.
2. The city will provide irrigation for the first 5 years at a cost of approximately $0.50 per tree annually.
3. Planting costs are $100 per tree for trees instead of $120 per tree.

To calculate the dollar value of total benefits for the 40-year period, values from the last columns in the benefit tables in appendix 2 (40-year average) are multiplied by 40 years. As this value is for one tree, it must be multiplied by the total number of trees planted in the respective small, medium, or large tree size classes. To adjust for lower electricity prices, we multiply electricity saved for each tree type in the RU column by the number of trees and 40 years (large tree: 162 kWh × 500 trees × 40 years = 3,240,000 kWh). This value is multiplied by the price of electricity in Mangrovia ($0.062/kWh × 3,240,000 kWh = $200,880) to obtain cumulative air-conditioning energy savings for the project (table 5).

All the benefits are summed for each size tree for a 40-year period. The 500 small trees provide $615,000 in total benefits. The medium and large trees provide approximately $1.3 million and $1.9 million, respectively.

Table 5—Spreadsheet calculations of benefits and costs for the Mangrovia planting project, comparison over 40 years

	500 small		500 medium		500 large	
	Resource units	Total value	Resource units	Total value	Resource units	Total value
		Dollars		*Dollars*		*Dollars*
Benefits:						
Electricity (kWh)	740,000	45,880	1,560,000	96,720	3,240,000	200,880
Net carbon dioxide (lb)	3,480,000	11,643	3,760,000	12,586	7,400,000	24,718
Ozone (lb)	3,250	4,781	6,100	8,981	12,000	17,674
Nitrogen dioxide (lb)	8,930	13,156	20,600	30,340	23,630	34,797
Sulfur dioxide (lb)	7,840	11,948	18,100	27,588	20,670	31,504
Small particulate matter (lb)	4,920	6,583	10,200	13,648	14,630	19,592
Volatile organic compounds (lb)	1,500	904	3,490	2,105	3,880	2,339
Biogenic volatile organic compounds (lb)	-270	-163	-410	-249	-380	-230
Hydrology (gal)	12,100,000	120,993	24,740,000	247,400	42,160,000	421,503
Aesthetics and other benefits		399,626		900,122		1,145,345
Total benefits		615,351		1,339,242		1,898,122
Costs:						
Tree and planting ($)		50,000		50,000		50,000
Pruning ($)		198,400		317,200		338,200
Remove and dispose ($)		20,400		18,600		28,000
Infrastructure ($)		110,600		101,400		150,200
Irrigation ($)		1,250		1,250		1,250
Cleanup ($)		36,800		33,800		50,000
Liability and legal ($)		0		0		0
Administration and other ($)		41,400		38,000		56,400
Total costs		458,850		560,250		674,050
Net benefits		156,501		778,992		1,224,072
Benefit/cost ratio		1.34		2.39		2.82

To adjust cost figures, we add a value for irrigation by multiplying the annual cost by the number of trees and by the number of years that irrigation will be applied ($0.50 × 500 trees × 5 years = $1,250). We multiply 500 large trees by the unit planting cost ($100) to obtain the adjusted cost for planting (500 × $100 = $50,000). The average annual 40-year costs taken from the cost tables in appendix 2 for other items are multiplied by 40 years and the number of trees to compute total costs. These 40-year cost values are entered into table 5. The total costs for the small, medium, and large trees are $458,850, $560,250, and $674,050.

Subtracting total costs from total benefits yields net benefits for the small ($156,501), medium ($778,992), and large ($1.2 million) trees (table 5). The net benefits per street tree over the 40-year period are as follows:

- $313 for a small tree
- $1,558 for a medium tree
- $2,448 for a large tree

When small trees are planted instead of large trees, the residents of Mangrovia stand to lose benefits of more than $2,000 per tree. In a new neighborhood with 500 trees, the total loss of benefits would exceed $1 million over the project lifetime.

Based on this analysis, the city of Mangrovia decided to develop and enforce street tree ordiance that requires planting large trees where possible and requires tree shade plans that show how developers will achieve 50 percent shade over streets, sidewalks, and parking lots within 15 years of development.

This analysis assumes that 19 percent of the planted trees died. It does not account for the time value of money from a capital investment perspective, but this could be done by using the municipal discount rate.

Increasing Program Cost-Effectiveness

What if the program you have designed looks promising in terms of stormwater-runoff reduction, energy savings, volunteer participation, and additional benefits, but the costs are too high? This section describes some steps to consider that may increase benefits and reduce costs, thereby increasing cost-effectiveness.

Increasing Benefits

Improved stewardship to increase the health and survival of recently planted trees is one strategy for increasing cost-effectiveness. An evaluation of the Sacramento Shade program found that tree survival rates had a substantial impact on projected benefits (Hildebrandt et al. 1996). Higher survival rates increase energy savings and reduce tree removal and planting costs.

Energy benefits can be further increased by planting a higher percentage of trees in locations that produce the greatest energy savings, such as opposite west-facing walls and close to buildings with air conditioning. By customizing tree locations to increase numbers in high-yield sites, energy savings can be boosted.

Conifers and broadleaf evergreens intercept rainfall and particulate matter year round as well as provide shade, which lowers cooling costs. Locating these types of trees in yards, parks, school grounds, and other open-space areas can increase benefits.

Reducing Program Costs

Cost effectiveness is influenced by program costs as well as benefits:

Cost effectiveness = total net benefit / total program cost

Cutting costs is one strategy to increase cost effectiveness. A substantial percentage of total program costs occur during the first 5 years and are associated with tree planting and establishment (McPherson 1993). Some strategies to reduce these costs include:

- Plant bare-root or smaller tree stock.
- Use trained volunteers for planting and pruning of young trees (fig. 19).
- Provide followup care to increase tree survival and reduce replacement costs.
- Select and locate trees to avoid conflicts with infrastructure.

Where growing conditions are likely to be favorable, such as yard or garden settings, it may be cost-effective to use smaller, less expensive stock or bare-root trees. In highly urbanized settings and sites subject to vandalism, however, large stock may survive the initial establishment period better than small stock.

Although organizing and training volunteers requires labor and resources, it is usually less costly than contracting the work, and it can help build more support for your program. A cadre of trained volunteers can easily maintain trees until they reach a height of about 20 ft and limbs are too high to prune from the ground with pole pruners. By the time trees reach this size, they are well established. Pruning during this establishment period should result in trees that will require less care in the long term. Training young trees can provide a strong branching structure that requires less frequent thinning and shaping (Costello 2000). Ideally, young trees should be inspected and pruned every other year for the first 5 years after planting.

Tree Trust

Figure 19—Trained volunteers can plant and maintain young trees, allowing the community to accomplish more at less cost and providing satisfaction for participants.

As trees grow larger, pruning costs may increase on a per-tree basis. The frequency of pruning will influence these costs, as it takes longer to prune a tree that has not been pruned in 10 years than one that was pruned a few years ago. Good structural pruning will improve performance in hurricanes (UF IFAS Extension 2006a).

Investing in the resources needed to promote tree establishment during the first 5 years after planting is usually worthwhile, because once trees are established they have a high probability of continued survival. If your program has targeted trees on private property, then encourage residents to attend tree-care workshops. These workshops should include information on recognizing pests and diseases and contact information for notifying authorities should an outbreak occur. Residents are an important first line of defense against pests in the Tropical region, which is especially prone to attacks from nonnative species. Develop standards of "establishment success" for different types of tree species. Perform periodic inspections to alert residents to tree health problems, and reward those whose trees meet your program's establishment standards. Replace dead trees as soon as possible, and identify ways to improve survivability.

Carefully select and locate trees to avoid conflicts with overhead power lines, sidewalks, and underground utilities. Time spent planning the planting will result in long-term savings. Also consider soil type and irrigation, microclimate, and the type of activities occurring around the tree that will influence its growth and management.

When evaluating the bottom line—trees pay us back—do not forget to consider benefits other than the stormwater-runoff reductions, energy savings, atmospheric CO_2 reductions, and other tangible benefits. The magnitude of benefits related to employment opportunities, job training, community building, reduced violence, and enhanced human health and well-being can be substantial (fig. 20). Moreover, these benefits extend beyond the site where trees are planted, furthering collaborative efforts to build better communities.

For more information on urban and community forestry program design and implementation, see the list of additional resources in appendix 1.

Figure 20—Trees pay us back in tangible and intangible ways.

Chapter 5. General Guidelines for Selecting and Placing Trees

Guidelines for Energy Savings
Maximizing Energy Savings From Shading

The right tree in the right place can save energy and reduce tree care costs. The sun shines on the east side of a building in the morning, passes over the roof near midday, and then shines on the west side in the afternoon. Electricity use for cooling is highest during the afternoon when temperatures are warmest and incoming sunshine is greatest. Therefore, the west side of a home is the most important side to shade (Sand 1993) (fig. 21). Depending on building orientation and window placement, sun shining through windows can heat a home quickly during the morning hours. The east side is the second most important side to shade when considering the net impact of tree shade on energy savings (fig. 21).

The closer a tree is to a home the more shade it provides, but roots of trees that are too close can damage the foundation. Branches too close to the building can make it difficult to maintain exterior walls and windows. In addition, trees with branches overhanging the roof will drop leaves and wood onto the roof. In Tropical communities in particular, where roofs are more likely to be flat, debris can accumulate and cause damage as it rots. Keep trees 10 ft or farther from the home depending on mature crown spread, to avoid these conflicts. Trees within 30 to 50 ft of the home most effectively shade windows and walls. In addition, trees with branches overhanging the roof will drop leaves and wood onto the roof.

Figure 21—Locate trees to shade west and east windows (from Sand 1993).

51

Paved patios and driveways can become **heat sinks** that warm the home during the day. Shade trees can make them cooler and more comfortable spaces. If a home is equipped with an air conditioner, shading can reduce its energy use, but do not plant vegetation so close that it will obstruct the flow of air around the unit.

Plant only small-growing trees under overhead power lines, and avoid planting directly above underground water and sewer lines if possible. Contact your local utility company before planting to determine where underground lines are located and which tree species should not be planted below power lines.

Selecting Trees to Maximize Benefits

The ideal shade tree has a fairly dense, round crown with limbs broad enough to partially shade the roof. Given the same placement, a large tree will provide more shade than a small tree. Plant small trees where nearby buildings or power lines limit aboveground space. Columnar trees are appropriate in narrow side yards. Because the best location for shade trees is relatively close to the west and east sides of buildings, the most suitable trees will be strong and capable of resisting storm damage, disease, and pests (Sand 1994).

When selecting trees, match the tree's water requirements with those of surrounding plants. Also, match the tree's maintenance requirements with the amount of care and the type of use different areas in the landscape receive. For instance, tree species that drop fruit that can be a slip-and-fall problem should not be planted near paved areas frequently used by pedestrians. Check with your local landscape professional before selecting trees to make sure that they are well suited to the site's soil and climatic conditions.

Use the following practices to plant and manage trees strategically to maximize energy conservation benefits:

- Increase community-wide tree canopy, and target shade to streets, parking lots, and other paved surfaces, as well as air-conditioned buildings.
- Shade west- and east-facing windows and walls.
- Shade air conditioners, but don't obstruct airflow.
- Avoid planting trees too close to utilities and buildings.

Guidelines for Reducing Carbon Dioxide

Because trees in common areas and other public places may not shelter buildings from sun and wind and reduce energy use, carbon dioxide (CO_2) reductions are primarily due to sequestration. Fast-growing trees sequester more CO_2 initially than slow-growing trees, but this advantage can be lost if the fast-growing trees die at younger ages. Large trees have the capacity to store more CO_2 than smaller trees. To maximize CO_2 sequestration, select tree species that are well suited to the site where they will be planted. Consult with your local arborist to select the right tree for your site. Trees that are not well adapted will grow slowly, show symptoms of stress, or die at an early age. Unhealthy trees do little to reduce atmospheric CO_2 and can be unsightly liabilities in the landscape.

Design and management guidelines that can increase CO_2 reductions include the following:

- Maximize use of woody plants, especially trees, as they store more CO_2 than do herbaceous plants and grasses.
- Plant more trees where feasible and immediately replace dead trees to compensate for CO_2 lost through removal.
- Create diverse habitats, with trees of different ages and species, to promote a continuous canopy cover over time.
- Group species with similar landscape maintenance requirements together and consider how irrigation, pruning, fertilization, weed, pest, and disease control can be minimized.
- Reduce CO_2 associated with landscape management by using push mowers (not gas or electric), hand saws (not chain saws), pruners (not gas/electric shears), rakes (not leaf blowers), and employ landscape professionals who don't have to travel far to your site.
- Reduce maintenance by reducing turfgrass and planting sustainable landscapes.
- Consider the project's lifespan when selecting species. Fast-growing species will sequester more CO_2 initially than slow-growing species, but may not live as long.
- Provide ample space belowground for tree roots to grow so that they can maximize CO_2 sequestration and tree longevity.
- When trees die or are removed, salvage as much wood as possible for use as furniture and other long-lasting products to delay decomposition or use as a bioenergy source.
- Plant trees, shrubs, and vines in strategic locations to maximize summer shade and reduce winter shade, thereby reducing atmospheric CO_2 emissions associated with power production.

Guidelines for Reducing Stormwater Runoff

Trees are mini-reservoirs, controlling runoff at the source because their leaves and branch surfaces intercept and store rainfall, thereby reducing runoff volumes and erosion of watercourses, as well as delaying the onset of peak flows. Rainfall interception by large trees is a relatively inexpensive first line of defense in the battle to control nonpoint-source pollution.

When selecting trees to maximize rainfall interception benefits, consider the following:

- Select tree species with physiological features that maximize interception, such as evergreen foliage, large leaf surface area, and rough surfaces that store water (Metro 2002).
- Increase interception by planting large trees where possible (fig. 22).
- Plant trees that are in leaf when precipitation levels are highest.

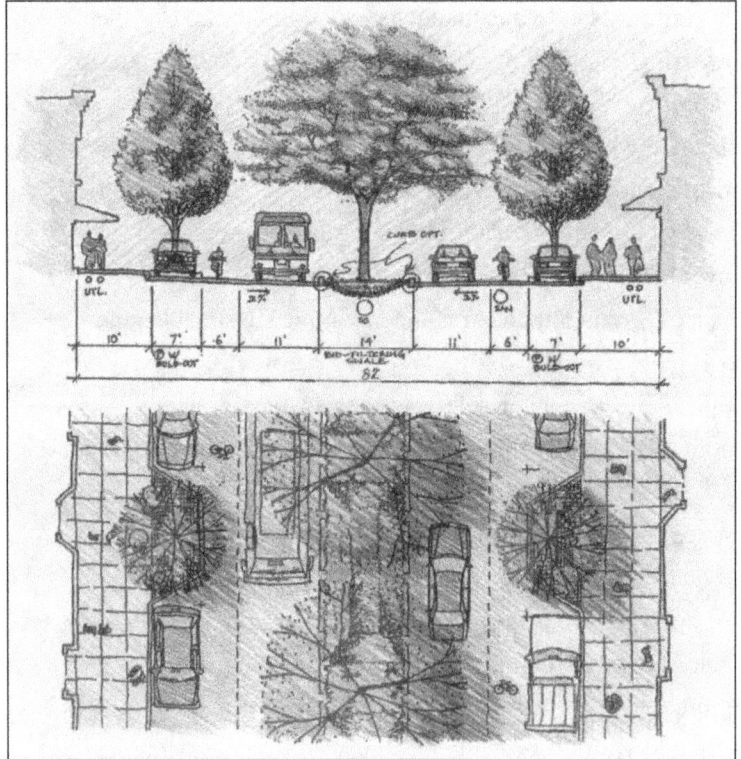

Figure 22—Trees can create a continuous canopy for maximum rainfall interception, even in commercial areas. In this example, a swale in the median filters runoff and provides ample space for large trees. Parking-space-sized planters contain the soil volume required to grow healthy, large trees (from Metro 2002).

- Plant low-water-use tree species where appropriate and native species that, once established, require little supplemental irrigation.
- In bioretention areas, such as roadside swales, select species that tolerate inundation, are long-lived, wide-spreading, and fast-growing (Metro 2002).
- Do not pave over streetside planting strips for easier weed control; this can impair tree health and increase runoff.
- Bioswales in parking lots and other paved areas store and filter stormwater while providing good conditions for trees.

Guidelines for Improving Air Quality Benefits

Trees, sometimes called the "lungs of our cities," are important because of their ability to remove contaminants from the air. The amount of gaseous pollutants and particulates removed by trees depends on their size and architecture, as well as local meteorology and pollutant concentrations.

Along streets, in parking lots, and in commercial areas, locate trees to maximize shade on paving and parked vehicles. Shade trees reduce heat that is stored or reflected by paved surfaces. By cooling streets and parking areas, trees reduce emissions of evaporative hydrocarbons from parked cars and thereby reduce smog formation (Scott et al. 1999). Large trees can shade a greater area than smaller trees, but should be used only where space permits. Remember that a tree needs space for both branches and roots. Keep in mind also that the soil along streets and parking lots will likely be compacted, and measures to reduce this problem, such as the use of engineered or structural soils, must be taken.

Tree planting and management guidelines to improve air quality include the following (Nowak 2000, Smith and Dochinger 1976):

- Broadleaf evergreens and conifers have high surface-to-volume ratios and retain their foliage year round, which may make them more effective than deciduous species.
- Species with long leaf stems and hairy plant parts are especially efficient interceptors.
- Effective uptake depends on proximity to the pollutant source and the amount of biomass. Where space and fire conditions permit, plant multilayered stands near the source of pollutants.
- In areas with unhealthy ozone (O_3) concentrations, maximize use of plants that emit low levels of biogenic volatile organic compounds to reduce O_3 formation—for example, members of the pea family.

- Sustain large, healthy trees; they produce the most benefits.
- To reduce emissions of volatile organic compounds and other pollutants, plant trees to shade parked cars and conserve energy.
- When possible, keep trees in mind in the early stages of planning new development, especially municipal and regional infrastructure plans. Trees can be included as air pollution mitigation measures along new traffic corridors.

Guidelines for Avoiding Conflicts With Infrastructure

Trees can become liabilities when they conflict with power lines, underground utilities, and other infrastructure elements. Guidelines to reduce conflicts with infrastructure include the following:

- Before planting, contact your local before-digging company, such as One Call or Call Before You Dig, to locate underground water, sewer, gas, and telecommunications lines.
- Avoid locating trees where they will block streetlights or views of traffic and commercial signs.
- Check with local transportation officials for sight visibility requirements. Keep trees at least 30 ft away from street intersections to ensure visibility.
- Avoid planting shallow-rooting species near sidewalks, curbs, and paving where tree roots can heave pavement if planted too close. Generally, avoid planting within 3 ft of pavement, and remember that trunk flare at the base of large trees can displace soil and paving for a considerable distance. Consider strategies to reduce damage by tree roots such as meandering sidewalks around trees (Costello and Jones 2003).
- Plant only small trees (<25 ft tall) under overhead power lines, and do not plant directly above underground water and sewer lines (fig. 23). Avoid locating trees where they will block illumination from streetlights or views of street signs in parking lots, commercial areas, and along streets.

Maintenance requirements and public safety concerns influence the type of trees selected for public places. The ideal public tree is not susceptible to wind damage and branch drop, does not require frequent pruning, produces negligible litter, is deep-rooted, has few serious pest and disease problems, and tolerates a wide range of soil conditions, irrigation regimes, and air pollutants. Because relatively few trees have all these traits, it is important to match the tree species to

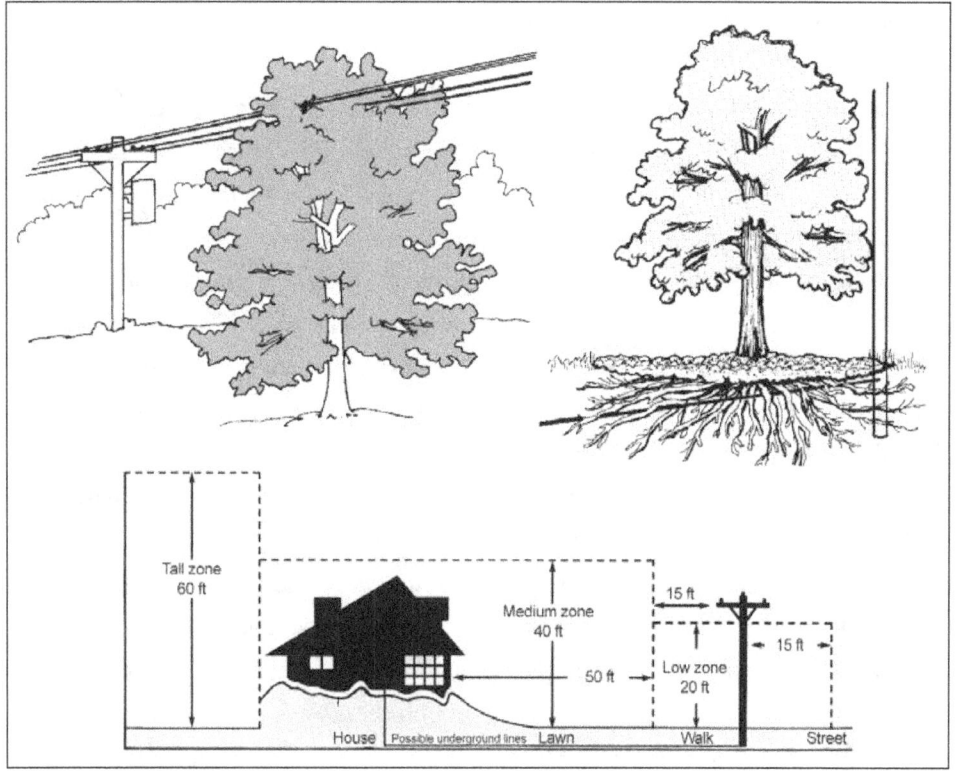

Figure 23—Know where power lines and other utility lines are before planting. Under power lines, use only small-growing trees ("low zone") and avoid planting directly above underground utilities. Larger trees may be planted where space permits ("medium" and "tall zones") (from ISA 1992).

the planting site by determining what issues are most important on a case-by-case basis. For example, parking-lot trees should be tolerant of hot, dry conditions, have strong branch attachments, and be resistant to attacks by pests that leave vehicles covered with sticky exudates. Check with your local landscape professional for horticultural information on tree traits.

Guidelines for Maximizing Long-Term Benefits

Invasive Nonnative Species

In the previous sections, we have offered suggestions for choosing trees to achieve certain goals. A basic underlying rule, however, must always be to choose species that are not, on balance, harmful. Special care should be taken when selecting plants to avoid those that are invasive nonnative species. Invasive species are plants that have been brought to a region for aesthetic or agricultural reasons or that have been introduced accidentally and that are able to gain a special foothold in their new environment. In Tropical communities, they can destroy native ecosystems, displace native plants, disturb habitats for native fauna, and increase fire risk (Smith 1998).

Not all nonnative plant species are dangerous; many have great economic or aesthetic value and pose little risk to their ecosystems. The difficulty lies in distinguishing between the two. In Hawaii, several organizations, including the University of Hawaii, the USDA Forest Service, the Hawaii Department of Land and Natural Resources, and the Hawaiian Ecosystems at Risk project, are working to identify dangerous nonnative species before they become pests. The Department of Land and Natural Resources (DNLR) publishes the list *Hawaii's Most Invasive Horticultural Plants* (DLNR 2007). A joint project by scientists at the University of Hawaii and the USDA Forest Service, called the Hawaii/Pacific Weed Risk Assessment, attempts to estimate the likeliness that a plant will become invasive based on published data (Daehler et al. 2004). These scientists are also beginning to make use of local information and local sources to improve predictions in a project called the Hawaii Exotic Plant Evaluation Protocol (Denslow and Daehler 2007). The Hawaiian Ecosystems at Risk project is a good source of information on the weed risk assessments and nonnative species in general (http://www.hear.org/).

In Florida, the Florida Exotic Plant Pest Council publishes an annual list of species known to have caused or likely to cause ecological damage (FLEPPC 2007). The FLEPPC Web site (http://www.fleppc.org) has many other helpful references, including links to individual county resources on prohibited plants, scientific and general publications to provide more information, and ways to get involved in helping to combat the problem of invasive species. In the Caribbean, Kairos et al. (2003) provided a helpful overview of invasive species threats, including plant lists.

Planting Guidelines

Selecting a tree from the nursery that has a high probability of becoming a healthy, trouble-free **mature tree** is critical to a successful outcome. Therefore, select the very best stock at your nursery, and when necessary, reject stock that does not meet industry standards. The University of Florida Institute of Food and Agricultural Sciences Extension (2006b) and the Urban Forest Ecosystem Institute (2008) provide a good starting point for communities interested in creating their own standards for nurseries or for assessing quality.

The health of the tree's root ball is critical to its ultimate survival. If the tree is in a container, check for matted roots by sliding off the container. Roots should penetrate to the edge of the root ball, but not densely circle the inside of the container or grow through drain holes. As well, at least two large structural roots should emerge from the trunk within 1 to 3 in of the soil surface. If there are no roots in the upper portion of the root ball, it is undersized and the tree should not be planted.

Another way to evaluate the quality of the tree before planting is to gently move the trunk back and forth. A good tree trunk bends and does not move in the soil, whereas a poor trunk bends a little and pivots at or below the soil line—a tell-tale sign of a poorly anchored tree.

Dig the planting hole 1 in shallower than the depth of the root ball to allow for some settling after watering. Make the hole two to three times as wide as the root ball and loosen the sides of the hole to make it easier for roots to penetrate. Place the tree so that the root flare is at the top of the soil. If the structural roots have grown properly as described above, the top of the root ball will be slightly higher (1 to 2 in) than the surrounding soil to allow for settling. Backfill with the native soil unless it is very rocky or sandy, in which case you may want to add composted organic matter such as peat moss or shredded bark (fig. 24).

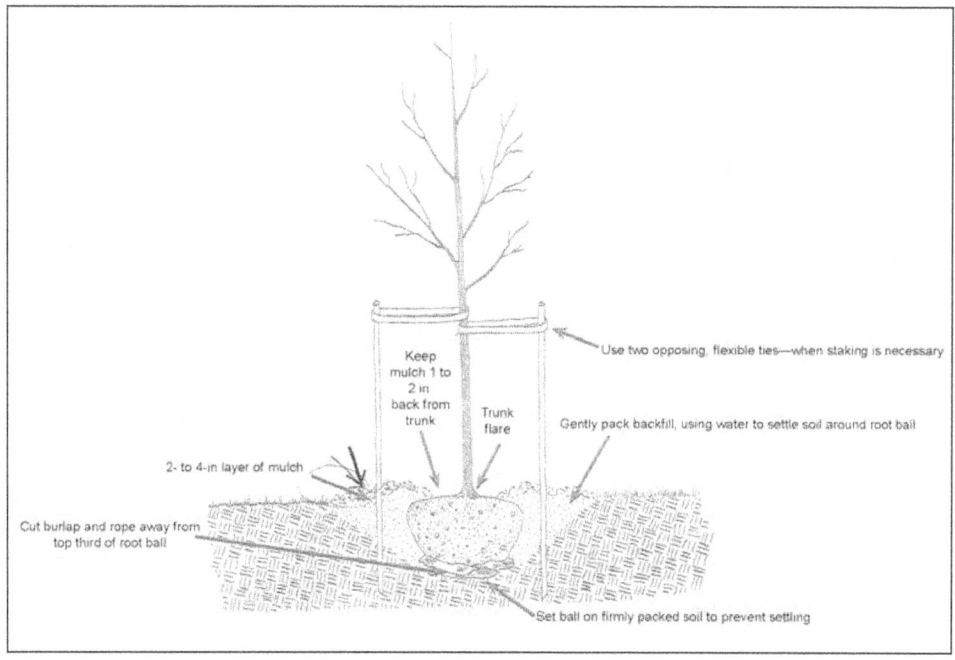

Figure 24—Prepare a broad planting area, plant the tree with the root flare at or just above ground level, and provide a berm/water ring to retain water (drawing courtesy of International Society of Arboriculture). (Note that trunk flare shown here represents a tree grown under optimum conditions. In trees grown under poorer conditions, the trunk flare may be hidden beneath the soil. These trees should be rejected in favor of those grown more carefully, or at the very least, the soil should be removed to expose the flare.)

Planting trees in urban plazas, commercial areas, and parking lots poses special challenges because of limited soil volume and poor soil structure. For trees to deliver benefits over the long term they require enough soil volume to grow and remain healthy. Matching tree species to the site's soil volume can reduce sidewalk and curb damage as well. Figure 25 shows recommended soil volumes for different sized trees. Engineered or structural soils can be placed under the hardscape to increase rooting space while meeting engineering requirements. For more information on structural soils see *Reducing Infrastructure Damage by Tree Roots: A Compendium of Strategies* (Costello and Jones 2003) and *Engineered Soil, Trees and Stormwater Runoff: The UC Davis Parking Lot Project* (CUFR 2007).

Use the extra soil left after planting to build a berm outside the root ball that is 6 in high and 3 ft in diameter. Soak the tree, and gently rock it to settle it in. Cover the basin with a 2- to 4-in layer of mulch, but avoid placing mulch against the tree trunk. Water the new tree two to three times a week and increase the amount of water as the tree grows larger. Generally, a tree requires about 1 in of water per week. A rain gauge or soil moisture sensor (tensiometer) can help determine tree watering needs, or contact your local cooperative extension agent or water conservancy district for recommendations.

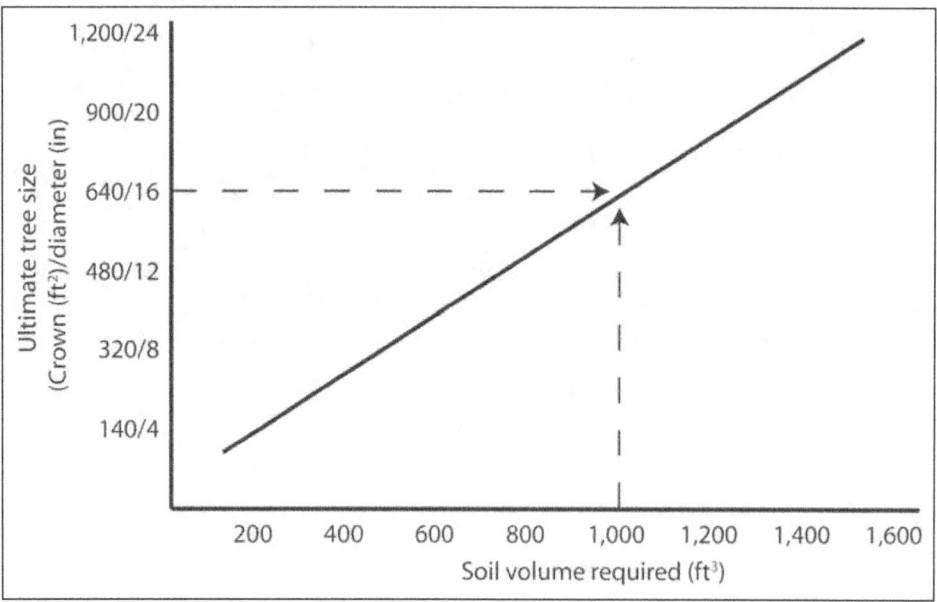

Figure 25—Developed from several sources by Urban (1992), this graph shows the relationship between tree size and required soil volume. For example, a tree with a 16-in diameter at breast height (41 cm) with 640 ft^2 of crown projection area (59.5 m^2 under the dripline) requires 1,000 ft^3 (28 m^3) of soil (from Costello and Jones 2003).

After you've planted your tree, remember the following:

- Inspect your tree several times a year, and contact a local landscape professional if problems develop.

- If your tree needed staking to keep it upright, remove the stake and ties after 1 year or as soon as the tree can hold itself up. The staking should allow some tree movement, as this movement sends hormones to the roots causing them to grow and create greater tree stability. It also promotes trunk taper and growth.

- Reapply mulch and irrigate the tree as needed.

- Leave lower side branches on young trees for the first year and prune back to 4 to 6 in to accelerate tree diameter development. Remove these lateral branches after the first full year. Prune the young tree to maintain a central main trunk and equally spaced branches. For more information, see Costello (2000) and Gilman (2002). As the tree matures, have it pruned on a regular basis by a certified arborist or other experienced professional.

By keeping your tree healthy, you maximize its ability to produce shade, intercept rainfall, reduce atmospheric CO_2, and provide other benefits. For more information on tree selection, planting, establishment, and care, see the resources listed in appendix 1.

Trees for Hurricane-Prone Areas

In addition to the damage they cause to urban infrastructure, hurricanes can also have a significant impact on a city's green infrastructure. Trees may be uprooted, snapped, or may lose large branches. But hurricanes don't affect all trees or all tree species equally. A study in Florida after several hurricanes between 1995 and 2005 showed that some species stood a better chance of surviving (Duryea et al. 2007a). The most wind-resistant tropical trees were live oak, sabal palm, and sand live oak (see "Common and Scientific Names" section). The least wind resistant were sand pine, Carolina laurelcherry, and the Washington fan palm. Another study in south Florida and Puerto Rico showed highest survival rates after hurricanes for bald cypress, camphor, gumbo limbo, Santa Maria, Caribbean pine, schefflera, and West Indian mahogany (Duryea et al. 2007b). Poorly surviving species included melaleuca, Australian pine, African tuliptree, and weeping banyan. With the exception of the Queen palm, palms showed excellent survival in both studies. Other studies have shown pond and bald cypress to be extremely wind resistant (Ogden 1992).

General suggestions for selecting and maintaining palms include:

- Preferred species have wide-spreading branches, strong, deep root systems, low centers of gravity, and small leaves.
- Work to increase the number of native trees planted—they have stood the test of time.
- Plant trees in masses wherever possible for least damage (Duryea 2007a, 2007b).
- Provide trees with plenty of room for roots to grow—they provide the anchor that holds the tree in place. Root pruning will greatly reduce stability.
- Prune trees appropriately to remove weak branches and improve structure. (See UF IFAS Extension 2006a for extensive information on pruning.) "Topping" and "lion-tailing" in particular reduce tree strength, raise the tree's center of gravity, and reduce its aerodynamics.
- Municipalities, utility companies, and others responsible for tree maintenance should draft standards for pruning and care to reduce hurricane-related damage.

Glossary of Terms

annual fuel utilization efficiency (AFUE)—A measure of space heating equipment efficiency defined as the fraction of energy output per energy input.

anthropogenic—Produced by humans.

biodiversity—The variety of life forms in a given area. Diversity can be categorized in terms of the number of species, the variety in the area's plant and animal communities, the genetic variability of the animals or plants, or a combination of these elements.

biogenic—Produced by living organisms.

biogenic volatile organic compounds (BVOCs)—Hydrocarbon compounds from vegetation (e.g., isoprene, monoterpene) that exist in the ambient air and contribute to the formation of smog or may themselves be toxic. Emission rates ($\mu g \cdot g^{-1} \cdot hr^{-1}$) used for this report follow Benjamin and Winer (1998):

- *Conocarpus erectus* var. *argenteus* —0.1 (isoprene); 0.2 (monoterpene)
- *Cassia x nealiae*—0.0 (isoprene); 0.2 (monoterpene)
- *Samanea saman*—0.0 (isoprene); 0.2 (monoterpene)

canopy—A layer or multiple layers of branches and foliage at the top or crown of a forest's trees.

canopy cover—The area of land surface that is covered by tree canopy, as seen from above.

Ccf—One hundred cubic feet.

climate—The average weather for a particular region and period (usually 30 years). Weather describes the short-term state of the atmosphere; climate is the average pattern of weather for a particular region. Climatic elements include precipitation, temperature, humidity, sunshine, wind velocity, phenomena such as fog, frost, and hailstorms, and other measures of weather.

climate effects—Impact on residential space heating and cooling (lb of carbon dioxide [CO_2] per tree per year) from trees located more than 50 ft from a building owing to associated reductions in windspeeds and summer air temperatures.

community forests—The sum of all woody and associated vegetation in and around human settlements, ranging from small rural villages to metropolitan regions.

contract rate—The percentage of residential trees cared for by commercial arborists; the proportion of trees for which a specific service (e.g., pruning or pest management) is contracted.

control costs—The marginal cost of preventing, controlling, or mitigating an impact.

crown—The branches and foliage at the top of a tree.

cultivar (derived from "*culti*vated *vari*ety")—Denotes certain cultivated plants that are clearly distinguishable from others by any characteristic, and that when reproduced (sexually or asexually), retain their distinguishing characteristics. In the United States, "variety" is often considered synonymous with "cultivar."

damage costs—The total estimated economic loss produced by an impact.

deciduous—Trees or shrubs that lose their leaves every fall.

diameter at breast height (d.b.h.)—The diameter of a tree outside the bark measured 4.5 ft above the ground on the uphill side (where applicable) of the tree.

dripline—The area beneath a tree marked by the outer edges of the branches.

emission factor—The rate of CO_2, nitrogen dioxide (NO_2), sulfur dioxide (SO_2), and small particulate matter (PM_{10}) output resulting from the consumption of electricity, natural gas, or any other fuel source.

evapotranspiration (ET)—The total loss of water by evaporation from the soil surface and by transpiration from plants, from a given area, and during a specified period.

evergreens—Trees or shrubs that are never entirely leafless. Evergreens may be broadleaved or coniferous (cone-bearing with needlelike leaves).

greenspace—Urban trees, forests, and associated vegetation in and around human settlements, ranging from small communities in rural settings to metropolitan regions.

hardscape—Paving and other impervious ground surfaces that reduce infiltration of water into the soil.

heat sinks—Paving, buildings, and other surfaces that store heat energy from the sun.

hourly pollutant dry deposition—Removal of gases from the atmosphere by direct transfer to natural surfaces and absorption of gases and particles by natural surfaces such as vegetation, soil, water, or snow.

interception—Amount of rainfall held on tree leaves and stem surfaces.

kWh (kilowatt-hour)—A unit of work or energy, measured as 1 kW (1,000 watts) of power expended for 1 hr.

leaf area index (LAI)—Total leaf area per unit area of crown if crown were projected in two dimensions.

leaf surface area (LSA)—Measurement of area of one side of a leaf or leaves.

mature tree—A tree that has reached a desired size or age for its intended use. Size, age, and economic maturity differ depending on the species, location, growing conditions, and intended use.

mature tree size—The approximate size of a tree 40 years after planting.

metric tonne—A measure of weight (abbreviated "t") equal to 1,000,000 grams (1,000 kilograms) or 2,205 pounds.

municipal forester—A person who manages public street and/or park trees (municipal forestry programs) for the benefit of the community.

MWh (megawatt-hour)—A unit of work or energy, measured as one megawatt (1,000,000 watts) of power expended for 1 hr. One MWh is equivalent to 3.412 MBtu.

nitrogen oxides (oxides of nitrogen, NO_x)—A general term for compounds of nitric oxide (NO), nitrogen dioxide (NO_2), and other oxides of nitrogen. Nitrogen oxides are typically created during combustion processes and are major contributors to smog formation and acid deposition. NO_2 may cause numerous adverse human health effects.

ozone (O_3)—A strong-smelling, pale blue, reactive toxic chemical gas consisting of three oxygen atoms. It is a product of the photochemical process involving the Sun's energy. Ozone exists in the upper layer of the atmosphere as well as at the Earth's surface. Ozone at the Earth's surface can cause numerous adverse human health effects. It is a major component of smog.

peak flow (or peak runoff)—The maximum rate of runoff at a given point or from a given area, during a specific period.

photosynthesis—The process in green plants of converting water and CO_2 into sugar with light energy; accompanied by the production of oxygen.

PM_{10} (particulate matter)—Major class of air pollutants consisting of tiny solid or liquid particles of soot, dust, smoke, fumes, and mists. The size of the particles (10 microns or smaller, about 0.0004 in or less) allows them to enter the air sacs (gas-exchange region) deep in the lungs where they may be deposited and cause adverse health effects. PM_{10} also reduces visibility.

reduced powerplant emissions—Reduced emissions of carbon dioxide (CO_2) or other pollutants that result from reductions in building energy use owing to the moderating effect of trees on climate. Reduced energy use for heating and cooling results in reduced demand for electrical energy, which translates into fewer emissions by powerplants.

resource unit (RU)—The value used to determine and calculate benefits and costs of individual trees. For example, the amount of air conditioning energy saved in kWh/year per tree, air-pollutant uptake in pounds per year per tree, or rainfall intercepted in gallons per tree per year.

riparian habitats—Narrow strips of land bordering creeks, rivers, lakes, or other bodies of water.

seasonal energy efficiency ratio (SEER)—Ratio of cooling output to power consumption; kBtu-output/kWh-input as a fraction. It is the Btu of cooling output during normal annual usage divided by the total electric energy input in kilowatt-hours during the same period.

sequestration—Annual net rate that a tree removes CO_2 from the atmosphere through the processes of photosynthesis and respiration (lb of CO_2 per tree per year).

shade coefficient—The percentage of light striking a tree crown that is transmitted through gaps in the crown. This is the percentage of light that hits the ground.

shade effects—Impact on residential space heating and cooling (lb of CO_2 per tree per year) from trees located within 50 ft of a building.

stem flow—Amount of rainfall that travels down the tree trunk and onto the ground.

sulfur dioxide (SO_2)—A strong-smelling, colorless gas that is formed by the combustion of fossil fuels. Powerplants, which may use coal or oil high in sulfur content, can be major sources of SO_2. Sulfur oxides contribute to the problem of acid deposition.

t—See metric tonne.

therm—A unit of heat equal to 100,000 British thermal units (BTUs) or 100 kBtu.

throughfall—Amount of rainfall that falls directly to the ground below the tree crown or drips onto the ground from branches and leaves.

transpiration—The loss of water vapor through the stomata of leaves.

tree or canopy cover—Within a specific area, the percentage covered by the crown of an individual tree or delimited by the vertical projection of its outermost perimeter; small openings in the crown are ignored. Used to express the relative importance of individual species within a vegetation community or to express the coverage of woody species.

tree litter—Fruit, leaves, twigs, and other debris shed by trees.

tree-related CO_2 emissions—CO_2 released when growing, planting, and caring for trees.

tree surface saturation storage capacity—The maximum volume of water that can be stored on a tree's leaves, stems, and bark. This part of rainfall stored on the canopy surface does not contribute to surface runoff during and after a rainfall event.

urban heat island—An area in a city where summertime air temperatures are 3 to 8 °F warmer than temperatures in the surrounding countryside. Urban areas are warmer for two reasons: (1) dark construction materials for roofs and asphalt absorb solar energy and (2) few trees, shrubs, or other vegetation provide shade and cool the air.

volatile organic compounds (VOCs)—Hydrocarbon compounds that exist in the ambient air. VOCs contribute to the formation of smog or are themselves toxic. VOCs often have an odor. Some examples of VOCs are gasoline, alcohol, and the solvents used in paints.

willingness to pay—The maximum amount of money an individual would be willing to pay for nonmarket, public goods and services provided by environmental amenities such as trees and forests rather than do without.

Common and Scientific Names[a]

Common name	Scientific name
African tuliptree	*Spathodea campanulata* P. Beauv.
Bald cypress	*Taxodium distichum* (L.) L.C. Rich.
Black oak	*Quercus velutina* Lam.
Camphor	*Cinnamomum camphora* (L.) J. Presl
Caribbean pine	*Pinus caribaea* Morelet
Carolina laurelcherry	*Prunus caroliniana* Ait.
Coconut palm	*Cocos nucifera* L.
Coral tree	*Erythrina* sp.
Gumbo limbo	*Bursera simaruba* (L.) Sarg.
Ironwood	*Casuarina equisetifolia* L.
Live oak	*Quercus virginiana* P. Mill.
Melaleuca	*Melaleuca quinquenervia* (Cav.) S.F. Blake
Monkeypod	*Samanea saman* (Jacq.) Merr.
Pond cypress	*Taxodium ascendens* Brongn.
Queen palm	*Syagrus romanzoffiana* (Cham.) Glassman
Rainbow shower tree	*Cassia* x *nealiae* H.S. Irwin & Barneby
Sabal palm	*Sabal palmetto* (Walt.) Lodd. ex J.A. & J.H. Schultes
Sand live oak	*Quercus geminata* Small
Sand pine	*Pinus clausa* (Chapman ex Engelm.) Vasey ex Sarg.
Santa Maria	*Calophyllum antillanum* Britt.
Schefflera	*Schefflera actinophylla* (Endl.) H.A.T. Harms
Silver buttonwood	*Conocarpus erectus* L. var. *argenteus* Millsp.
Washington fan palm	*Washingtonia robusta* H. Wendl.
Weeping banyan	*Ficus benjamina* L.
West Indian mahogany	*Swietenia mahogoni* (L.) Jacq.

[a] This list provides the scientific names for species mentioned in the text. It is not intended to serve as a list of recommended plants for the region.

Metric Equivalents

When you know:	Multiply by:	To find:
Inches (in)	25.4	Millimeters
Inches (in)	2.54	Centimeters
Feet (ft)	.305	Meters
Square feet (ft^2)	.0929	Square meters
Cubic feet (ft^3)	.0283	Cubic meters
Miles (mi)	1.61	Kilometers
Acres (ac)	.405	Hectares
Square miles (mi^2)	2.59	Square kilometers
Gallons (gal)	.00378	Cubic meters
Pounds (lb)	.454	Kilograms
Pounds per square foot (lb/ft^2)	4.882	Kilograms per square meter
Tons (ton)	.907	Metric tonnes
Therms	29.31	Kilowatt hours
Therms	.02931	Megawatt hours
Fahrenheit (°F)	(F - 32)0.55	Celsius

Acknowledgments

We greatly appreciate the assistance provided by Stan Oka and his staff: Brandon Au, June Harada, Clark Leavitt, and Clifford Umebayashi, with a special thanks to Terri-Ann Koike (Division of Urban Forestry, Department of Parks and Recreation, City and County of Honolulu, Hawaii); Joshlyn Sand (Honolulu Botanical Gardens, Division of Urban Forestry, Department of Parks and Recreation, City and County of Honolulu, Hawaii); Kevin Eckert (President, Arbor Global); Abner Undan (President, Trees of Hawaii); Lester Inouye (President, Lester H. Inouye and Associates); Ken Schmidt (Honolulu Land Information System, Department of Planning and Permitting, City and County of Honolulu, Hawaii); Robert Magota (Real Property Assessment Division, Department of Budget and Fiscal Services, City and County of Honolulu, Hawaii); Thomas Cole (Institute of Pacific Islands Forestry, Pacific Southwest Research Station, USDA Forest Service); Thomas J. Brandeis (Forest Inventory and Analysis, Southern Research Station, USDA Forest Service); Greg Ina, Scott Maco, Jim Jenkins, Aren Dottenwhy, and Karen Wise (Davey Resource Group); and Stephanie Huang and Aywon-Anh Nguyen (Center for Urban Forest Research).

We also thank Joseph Boscaglia (City of Naples, Florida), Michael Greenstein (City of Lantana, Florida), John Harris (Earth Advisors), Carol Kwan (Carol Kwan Consulting), Mark Leon (Sunshine Landscape), Sergio Vasquez (Tropical Tree

Services), Russell Komori (Iroquois Point Island Club), and Michael Kraus (Tree Works) for their assistance in providing information on tree care costs.

Carmen Hernández (Puerto Rico Forest Service Bureau), Dudley Hartel (Urban Forestry South, USDA Forest Service), Francisco Escobedo (University of Florida), Janet Thebaud Gillmar (University of Hawaii), and Charles Marcus (Florida Division of Forestry) provided helpful reviews of this work.

Keith Cline (USDA Forest Service, State and Private Forestry) and Teresa Trueman-Madriaga (Kaulunani Urban Forestry Program of the Hawaii Department of Land and Natural Resources, Division of Forestry and Wildlife and the USDA Forest Service) provided invaluable support for this project.

References

Akbari, H.; Davis, S.; Dorsano, S.; Huang, J.; Winnett, S., eds. 1992. Cooling our communities: a guidebook on tree planting and light-colored surfacing. Washington, DC: U.S. Environmental Protection Agency. 26 p.

Alliance for Community Trees. 2006. Tree by tree, street by street. http://actrees.org. (27 January 2007).

American Forests. 2005. Urban ecosystem analysis, city of Jacksonville, Florida: calculating the value of nature. Washington, DC. 12 p.

American Forests. 2007. Urban ecosystem analysis, Palm Beach County, Florida: calculating the value of nature. Washington, DC. 16 p.

Anderson, L.M.; Cordell, H.K. 1988. Residential property values improve by landscaping with trees. Southern Journal of Applied Forestry. 9: 162–166.

Bedker, P.J.; O'Brien, J.G.; Mielke, M.E. 1995. How to prune trees. NA-FR-01-95. [Newtown Square, PA]: U.S. Department of Agriculture, Forest Service, Northeastern Area State and Private Forestry. 28 p.

Benjamin, M.T.; Winer, A.M. 1998. Estimating the ozone-forming potential of urban trees and shrubs. Atmospheric Environment. 32: 53–68.

Bernhardt, E.; Swiecki, T.J. 1993. The state of urban forestry in California: results of the 1992 California Urban Forest Survey. Sacramento: California Department of Forestry and Fire Protection. 51 p.

Bratkovich, S.M. 2001. Utilizing municipal trees: ideas from across the country. NA-TP-06-01. Newtown Square, PA: U.S. Department of Agriculture, Forest Service, Northeastern Area State and Private Forestry. 91 p.

Brenzel, K.N., ed. 2001. Sunset western garden book. 7th ed. Menlo Park, CA: Sunset Books, Inc. 768 p.

Cappiella, K.; Schueler, T.; Wright, T. 2005. Urban watershed forestry manual. Ellicott City, MD: Center for Watershed Protection. 138 p.

Cardelino, C.A.; Chameides, W.L. 1990. Natural hydrocarbons, urbanization, and urban ozone. Journal of Geophysical Research. 95(9): 13971–13979

Center for Urban Forest Research [CUFR]. 2007. Engineered soil, trees and stormwater runoff: the UC Davis parking lot project. http://www.fs.fed.us/psw/programs/cufr/products/psw_cufr686_UCDParkingLot.pdf. (14 September 2007).

Chameides W.L.; Lindsay, R.W.; Richardson, J.; Kiang, C.S. 1988. The role of biogenic hydrocarbons in urban photochemical smog: Atlanta as a case study. Science. 241: 1473–1475.

Chave, J.; Andalo, C.; Brown, S.; Cairns, M.A.; Chambers, J.Q.; Eamus, D.; Folster, H.; Fromard, F.; Higuchi, N.; Kira, T.; Lescure, J.-P.; Nelson, B.W.; Ogawa, H.; Puig, H.; Riera, B.; Yamakura, T. 2005. Tree allometry and improved estimation of carbon stocks and balance in tropical forests. Oecologia. 145: 87–99.

Chicago Climate Exchange. 2007. About CCX. http://www.chicagoclimatex.com/about/index.html. (21 January 2007).

City and County of Honolulu. 2000. Runoff: E. Storm water quality design. Honolulu, HI. [Pages unknown.]

Cook, D.I. 1978. Trees, solid barriers, and combinations: alternatives for noise control. In: Proceedings of the national urban forestry conference. ESF Publ. 80-003. Syracuse, NY: State University of New York: 330–334.

Costello, L.R. 2000. Training young trees for structure and form. [Video] V99-A. Oakland, CA: University of California, Agriculture and Natural Resources, Communication Services Cooperative Extension Service.

Costello, L.R.; Jones, K.S. 2003. Reducing infrastructure damage by tree roots: a compendium of strategies. Cohasset, CA: Western Chapter of the International Society of Arboriculture. 119 p.

Daehler, C.C.; Denslow, J.S.; Ansari, S.; Kuo, H. 2004. A risk assessment system for screening out invasive pest plants from Hawaii and other Pacific Islands. Conservation Biology. 18: 360–368.

Denslow, J.S.; Daehler, C. 2007. Evaluation of exotic plants in Hawaii: Where we are. Where we could go. http://www.botany.hawaii.edu/faculty/daehler/WRA/hepep.pdf. (30 July 2007).

Department of Land and Natural Resources [DLNR]. 2007. Hawaii's most invasive horticultural plants. http://www.state.hi.us/dlnr/dofaw/hortweeds/specieslist.htm. (29 July 2007.)

Duryea, M.L.; Kampf, E.; Littell, R.C. 2007a. Hurricanes and the urban forest: I. Effects of Southeastern United States coastal plain tree species. Arboriculture and Urban Forestry. 33: 83–97.

Duryea, M.L.; Kampf, E.; Littell, R.C. 2007b. Hurricanes and the urban forest: II. Effects of tropical and subtropical tree species. Arboriculture and Urban Forestry. 33: 98–112.

Dwyer, J.F.; McPherson, E.G.; Schroeder, H.W.; Rowntree, R.A. 1992. Assessing the benefits and costs of the urban forest. Journal of Arboriculture. 18(5): 227–234.

Dwyer, M.C.; Miller, R.W. 1999. Using GIS to assess urban tree canopy benefits and surrounding greenspace distributions. Journal of Arboriculture. 25(2): 102–107.

Elliott, D. 2004. Hawaii wind mapping and validation project. Presentation at the Hawaii Wind Working Group Meeting, Honolulu, Hawaii, 6 January 2004. http://www.hawaii.gov/dbedt/ert/wwg/elliott040106.pdf. (20 June 2007).

European Climate Exchange. 2006. Historic data—ECX CFI futures contract. http://www.europeanclimateexchange.com. (29 August 2006).

Fazio, J.R. [N.d.]. Tree City USA Bulletin Series. Lincoln, NE: The National Arbor Day Foundation.

Florida Exotic Plant Pest Council [FLEPPC]. 2007. 2007 list of invasive plant species. http://www.fleppc.org/list/07list_brochure.pdf. (27 July 2007).

Frangi, J.L.; Lugo, A.E. 1985. Ecosystem dynamics of a subtropical floodplain forest. Ecological Monographs. 55(3): 351–369.

Freddie Mac. 2006. Making home possible in Puerto Rico. http://www.freddiemac.com/corporate/about/pdf/Puerto_Rico.pdf. (25 March 2008).

Geiger, J. 2001. Save dollars with shade. Davis, CA: U.S. Department of Agriculture, Forest Service, Pacific Southwest Research Station. 4 p. http://www.fs.fed.us/psw/programs/cufr/products/3/cufr_149.pdf. (25 January 2007).

Geiger, J. 2002a. Green plants or powerplants? Davis, CA: U.S. Department of Agriculture, Forest Service, Pacific Southwest Research Station. 4 p. http://www.fs.fed.us/psw/programs/cufr/products/3/cufr_148.pdf. (25 January 2007).

Geiger, J. 2002b. Where are all the cool parking lots? Davis, CA: U.S. Department of Agriculture, Forest Service, Pacific Southwest Research Station. 4 p. http://www.fs.fed.us/psw/programs/cufr/products/3/cufr_151.pdf. (25 January 2007).

Geiger, J. 2003. Is all your rain going down the drain? Davis, CA: U.S. Department of Agriculture, Forest Service, Pacific Southwest Research Station. 4 p. http://www.fs.fed.us/psw/programs/cufr/products/cufr_392_rain_down_the_drain.pdf. (25 January 2007).

Geiger, J. 2006. Trees—the air pollution solution. Davis, CA: U.S. Department of Agriculture, Forest Service, Pacific Southwest Research Station. 4 p. http://www.fs.fed.us/psw/programs/cufr/products/3/cufr_658_Air_pollution_solution.pdf. (25 January 2007).

Gilman, E.F. 1997. Trees for urban and suburban planting. Albany, NY: Delmar Publishing. 688 p.

Gilman, E.F. 2002. An illustrated guide to pruning. 2nd ed. Albany, NY: Delmar Publishing. 256 p.

Gonzalez, S. 2004. Personal communication. Assistant maintenance superintendent, City of Vallejo, 111 Amador St., Vallejo, CA 94590.

Grant, R.H.; Heisler, G.M.; Goa, W. 2002. Estimation of pedestrian level UV exposure under trees. Photochemistry and Photobiology. 75(4): 369–376.

Guenther, A.B.; Monson, R.K.; Fall, R. 1991. Isoprene and monoterpene emission rate variability: observations with eucalyptus and emission rate algorithm development. Journal of Geophysical Research. 96: 10799–10808.

Guenther, A.B.; Zimmermann, P.R.; Harley, P.C.; Monson, R.K.; Fall, R. 1993. Isoprene and monoterpene emission rate variability: model evaluations and sensitivity analyses. Journal of Geophysical Research. 98: 12609–12617.

73

Hammer, T.T.; Coughlin, R.; Horn, E. 1974. The effect of a large urban park on real estate value. Journal of the American Institute of Planning. July: 274–275.

Hammond J.; Zanetto, J.; Adams, C. 1980. Planning solar neighborhoods. Sacramento, CA: California Energy Commission. 179 p.

Hargrave, R.; Johnson, G.R.; Zins, M.E. 2002. Planting trees and shrubs for long-term health. MI-07681-S. St. Paul, MN: University of Minnesota Extension Service. 12 p.

Harris, R.W.; Clark, J.R.; Matheny, N.P. 2003. Arboriculture. 4th ed. Englewood Cliffs, NJ: Regents/Prentice Hall. 592 p.

Hightshoe, G.L. 1988. Native trees, shrubs, and vines for urban and rural America. New York: Van Nostrand Reinhold. 832 p.

Hildebrandt, E.W.; Kallett, R.; Sarkovich, M.; Sequest, R. 1996. Maximizing the energy benefits of urban forestation. In: Proceedings of the ACEEE 1996 summer study on energy efficiency in buildings. Washington, DC: American Council for an Energy Efficient Economy: 121–131. Vol. 9.

Hoyt, R.S. 1998. Ornamental plants for subtropical regions. Anaheim, CA: Livingston Press. 510 p.

Hudson, B. 1983. Private sector business analogies applied in urban forestry. Journal of Arboriculture. 9(10): 253–258.

Hull, R.B. 1992. How the public values urban forests. Journal of Arboriculture. 18(2): 98–101.

Intergovernmental Panel on Climate Change [IPCC]. 2001. Technical summary. climate change 2001: the scientific basis. Geneva. 63 p.

International Society of Arboriculture [ISA]. 1992. Avoiding tree and utility conflicts. Savoy, IL. 4 p.

International Society of Arboriculture. 2006. Welcome to the International Society of Arboriculture. http://www.isa-arbor.com/home.aspx. (27 January 2007).

Jo, H.K.; McPherson, E.G. 1995. Carbon storage and flux in residential greenspace. Journal of Environmental Management. 45: 109–133.

Kairos, M.; Ali, B.; Cheesman, O.; Haysom, K.; Murphy, S. 2003. Invasive species threats in the Caribbean region. London: CAB International. 134 p.

Kaplan, R. 1992. Urban forestry and the workplace. In: Gobster, P.H., ed. Managing urban and high-use recreation settings. St. Paul, MN: Gen. Tech. Rep. NC-163. U.S. Department of Agriculture, Forest Service, North Central Research Station: 41–45.

Kaplan, R.; Kaplan, S. 1989. The experience of nature: a psychological perspective. Cambridge, United Kingdom: Cambridge University Press. 360 p.

Kim, S.; Byun, D.W.; Cheng, F.-Y.; Czader, B.; Stetson, S.; Nowak, D.; Walton, J.; Estes, M.; Hitchcock, D. 2005. Modeling effects of land use land cover changes on meteorology and air quality in Houston, Texas, over the last two decades. In: Proceedings of the 2005 atmospheric sciences air quality conference. [Place of publication unknown]. http://ams.confex.com/ams/pdfpapers/92244.pdf. (23 January 2007).

Koike, T.-A. 2006. Personal communication. Administrative Assistant, Division of Urban Forestry, Department of Parks and Recreation, City and County of Honolulu, 3902 Paki Avenue, Honolulu, HI 96815.

Lewis, C.A. 1996. Green nature/human nature: the meaning of plants in our lives. Chicago: University of Illinois Press. 176 p.

Luley, C.J.; Bond, J. 2002. A plan to integrate management of urban trees into air quality planning. Naples, NY: Davey Resource Group. 61 p.

Maco, S.E.; McPherson, E.G. 2003. A practical approach to assessing structure, function, and value of street tree populations in small communities. Journal of Arboriculture. 29(2): 84–97.

Magota, R.O. 2007. Personal communication. Real property appraiser, City and County of Honolulu, Hawaii, 842 Bethel St., Honolulu, HI, 96813.

Mahoney, M.T.; Remyn, A.H.; Trotter, M.P.; Trotter, W.R.; Chamness, M.E.; Greby, K.J. 2000. Street trees recommended for southern California. Anaheim, CA: Street Tree Seminar. 200 p.

Marion, W.; Urban, K. 1995. User's manual for TMY2s—typical meteorological years. Golden, CO: National Renewable Energy Laboratory. 49 p.

Markwardt, L.J. 1930. Comparative strength properties of woods grown in the United States. Tech. Bull. 158. Washington, DC: U.S. Department of Agriculture. 38 p.

McHale, M.; McPherson, E.G.; Burke, I.C. 2007. The potential of urban tree plantings to be cost effective in carbon credit markets. Urban Forestry and Urban Greening. 6: 49–60.

McPherson, E.G. 1984. Planting design for solar control. In: McPherson, E.G., ed. Energy-conserving site design. Washington, DC: American Society of Landscape Architects: 141–164. Chapter 8.

McPherson, E.G. 1992. Accounting for benefits and costs of urban greenspace. Landscape and Urban Planning. 22: 41–51.

McPherson, E.G. 1993. Evaluating the cost effectiveness of shade trees for demand-side management. The Electricity Journal. 6(9): 57–65.

McPherson, E.G. 1995. Net benefits of healthy and productive forests. In: Bradley, G.A., ed. Urban forest landscapes: integrating multidisciplinary perspectives. Seattle, WA: University of Washington Press: 180–194.

McPherson, E.G. 2000. Expenditures associated with conflicts between street tree root growth and hardscape in California. Journal of Arboriculture. 26(6): 289–297.

McPherson, E.G. 2001. Sacramento's parking lot shading ordinance: environmental and economic costs of compliance. Landscape and Urban Planning. 57: 105–123.

McPherson, E.G.; Mathis, S., eds. 1999. Proceedings of the Best of the West summit. Sacramento, CA: Western Chapter, International Society of Arboriculture. 93 p.

McPherson, E.G.; Muchnick, J. 2005. Effects of tree shade on asphalt concrete pavement performance. Journal of Arboriculture. 31(6): 303–309.

McPherson, E.G.; Nowak, D.J.; Heisler, G.; Grimmond, S.; Souch, C.; Grant, R.; Rowntree, R.A. 1997. Quantifying urban forest structure, function, and value: Chicago's Urban Forest Climate Project. Urban Ecosystems. 1: 49–61.

McPherson, E.G.; Nowak, D.J.; Rowntree, R.A. 1994. Chicago's urban forest ecosystem: results of the Chicago Urban Forest Climate Project. Gen. Tech. Rep. NE-186. Radnor [Newtown Square], PA: U.S. Department of Agriculture, Forest Service, Northeastern Research Station. 201 p.

McPherson, E.G.; Peper, P.J. 1995. Infrastructure repair costs associated with street trees in 15 cities. In: Watson, G.W.; Neely, D., eds. Trees and building sites. Champaign, IL: International Society of Arboriculture: 49–63.

McPherson, E.G.; Sacamano, P.L.; Wensman, S. 1993. Modeling benefits and costs of community tree plantings. Albany, CA: U.S. Department of Agriculture, Forest Service, Pacific Southwest Research Station. 170 p.

McPherson, E.G.; Simpson, J.R. 1999. Carbon dioxide reduction through urban forestry: guidelines for professional and volunteer tree planters. Gen. Tech. Rep. PSW-171. Albany, CA: U.S. Department of Agriculture, Forest Service, Pacific Southwest Research Station. 237 p.

McPherson, E.G., Simpson, J.R. 2002. A comparison of municipal forest benefits and costs in Modesto and Santa Monica, CA, USA. Urban Forestry and Urban Greening. 1: 61–74.

McPherson, E.G.; Simpson, J.R. 2003. Potential energy savings in buildings by an urban tree planting programme in California. Urban Forestry and Urban Greening. 2(2): 73–86.

McPherson, E.G.; Simpson, J.R.; Peper, P.J.; Gardner, S.L.; Vargas, K.E.; Maco, S.E.; Xiao, Q. 2006a. Coastal Plain community tree guide: benefits, costs, and strategic planting. Gen Tech. Rep. PSW-GTR-201. Albany, CA: U.S. Department of Agriculture, Forest Service, Pacific Southwest Research Station. 99 p.

McPherson, E.G.; Simpson, J.R.; Peper, P.J.; Gardner, S.L.; Vargas, K.E.; Maco, S.E.; Xiao, Q. 2006b. Piedmont community tree guide: benefits, costs, and strategic planting. Gen Tech. Rep. PSW-GTR-200. Albany, CA: U.S. Department of Agriculture, Forest Service, Pacific Southwest Research Station. 95 p.

McPherson, E.G.; Simpson, J.R.; Peper, P.J.; Gardner, S.L.; Vargas, K.E.; Xiao, Q. 2007. Northeast community tree guide: benefits, costs, and strategic planting. Gen Tech. Rep. PSW-GTR-202. Albany, CA: U.S. Department of Agriculture, Forest Service, Pacific Southwest Research Station. 106 p.

McPherson, E.G.; Simpson, J.R.; Peper, P.J.; Maco, S.E.; Gardner, S.L.; Cozad, S.K.; Xiao, Q. 2006c. Midwest community tree guide: benefits, costs, and strategic planting. Gen. Tech. Rep. PSW-GTR-199. Albany, CA: U.S. Department of Agriculture, Forest Service, Pacific Southwest Research Station. 85 p.

McPherson, E.G.; Simpson, J.R.; Peper, P.J.; Maco, S.E.; Xiao, Q.; Hoefer, P.J. 2003. Northern mountain and prairie community tree guide: benefits, costs, and strategic planting. Albany, CA: U.S. Department of Agriculture, Forest Service, Pacific Southwest Research Station. 88 p.

McPherson, E.G.; Simpson, J.R.; Peper, P.J.; Maco, S.E.; Xiao, Q.; Mulrean, E. 2004. Desert southwest community tree guide: benefits, costs, and strategic planting. Albany, CA: U.S. Department of Agriculture, Forest Service, Pacific Southwest Research Station. 65 p.

McPherson, E.G.; Simpson, J.R.; Peper, P.J.; Scott, K.; Xiao, Q. 2000. Tree guidelines for coastal southern California communities. Sacramento, CA: Local Government Commission. 97 p.

McPherson, E.G.; Simpson, J.R.; Peper, P.J.; Xiao, Q. 1999a. Benefit-cost analysis of Modesto's municipal urban forest. Journal of Arboriculture. 25(5): 235–248.

McPherson, E.G.; Simpson, J.R.; Peper, P.J.; Xiao, Q. 1999b. Tree guidelines for San Joaquin Valley communities. Sacramento, CA: Local Government Commission. 63 p.

Meehl, G.A.; Washington, W.M.; Collins, W.D.; Arblaster, J.M.; Hu, A.; Buja, L.E.; Strand, W.G.; Teng, H. 2005. How much more global warming and sea level rise? Science. 307: 1769–1772.

Metro. 2002. Green streets: innovative solutions for stormwater and stream crossings. Portland, OR: Metro. 144 p.

Miller, R.W. 1997. Urban forestry: planning and managing urban greenspaces. 2^d ed. Upper Saddle River, NJ: Prentice-Hall. 502 p.

More, T.A.; Stevens, T.; Allen, P.G. 1988. Valuation of urban parks. Landscape and Urban Planning. 15: 139–152.

Morgan, R. [N.d.]. An introductory guide to community and urban forestry in Washington, Oregon, and California. Portland, OR: World Forestry Center. 25 p.

Morgan, R. 1993. A technical guide to urban and community forestry. Portland, OR: World Forestry Center. 49 p.

National Arbor Day Foundation. 2006. Color your world. http://www.arborday.org. (27 January 2007).

National Association of Realtors. 2007. Metropolitan area existing-home prices. http://www.realtor.org/Research.nsf/Pages/MetroPrice. (25 March 2008).

National Climatic Data Center. 2006. Quality controlled local climatological data. http://cdo.ncdc.noaa.gov/ulcdsw/ULCD. (21 November 2006).

Naughton, E. 2006. Personal communication. GIS Researcher, Hawaii Institute of Marine Biology, P.O. Box 1346, Kane'ohe, HI 96744.

Neely, D., ed. 1988. Valuation of landscape trees, shrubs, and other plants. 7th ed. Urbana, IL: International Society of Arboriculture. 50 p.

Nowak, D.J. 2000. Tree species selection, design, and management to improve air quality. In: Scheu, D.L., ed. 2000 ASLA annual meeting proceedings. Washington, DC: American Society of Landscape Architects: 23–27.

Nowak, D.J.; Civerolo, K.L.; Rao, S.T.; Sistla, G.; Luley, C.J.; Crane, D.E. 2000. A modeling study of the impact of urban trees on ozone. Atmospheric Environment. 34: 1601–1613.

Nowak, D.J.; Crane, D.E. 2002. Carbon storage and sequestration by urban trees in the USA. Environmental Pollution. 116: 381–389.

Ogden, J.C. 1992. The impact of Hurricane Andrew on the ecosystems of south Florida. Conservation Biology. 6: 488–490.

Oka, S. 2006. Personal communication. Urban forester, Urban Forestry Division, City and County of Honolulu, 3902 Paki Avenue, Honolulu, HI 96815.

Ottinger, R.L.; Wooley, D.R.; Robinson, N.A.; Hodas, D.R.; Babb, S.E. 1990. Environmental costs of electricity. New York: Oceana Publications. 769 p.

Parsons, R.; Tassinary, L.G.; Ulrich, R.S.; Hebl, M.R.; Grossman-Alexander, M. 1998. The view from the road: implications for stress recovery and immunization. Journal of Environmental Psychology. 18(2): 113–140.

Pearce, D. 2003. The social cost of carbon and its policy implications. Oxford Review of Public Policy. 19(3): 362–384.

Peper, P.J.; McPherson, E.G. 2003. Evaluation of four methods for estimating leaf area of isolated trees. Urban Forestry and Urban Greening. 2(1): 19–29.

Pillsbury, N.H.; Reimer, J.L.; Thompson, R.P. 1998. Tree volume equations for fifteen urban species in California. Tech. Rep. 7. San Luis Obispo, CA: Urban Forest Ecosystems Institute, California Polytechnic State University. 56 p.

Platt, R.H.; Rowntree, R.A.; Muick, P.C., eds. 1994. The ecological city. Boston, MA: University of Massachusetts. 292 p.

Pokorny, J.D. 2003. Urban tree risk management: a community guide to program design and implementation. NA-TP-03-03. Newtown Square, PA: U.S. Department of Agriculture, Forest Service, Northeastern Area State and Private Forestry. [Pages unknown].

Richards, N.A.; Mallette, J.R.; Simpson, R.J.; Macie, E.A. 1984. Residential greenspace and vegetation in a mature city: Syracuse, New York. Urban Ecology. 8: 99–125.

Rosenzweig, C.; Solecki, W.; Parshall, L.; Gaffin, S.; Lynn, B.; Goldberg, R.; Cox, J.; Hodges, S. 2006. Mitigating New York City's heat island with urban forestry, living roofs, and light surfaces. Presentation at 86th American Meteorological Society annual meeting, Jan. 31, 2006, Atlanta, Georgia. http://ams.confex.com/ams/pdfpapers/103341.pdf. (14 May 2008).

Sand, J. 2006. Personal communication. Horticulturist, Honolulu Botanical Gardens, 50 N Vineyard Blvd., Honolulu, HI 96817.

Sand, M. 1993. Energy saving landscapes: the Minnesota homeowner's guide. Minneapolis, MN: Minnesota Department of Natural Resources. [Pages unknown].

Sand, M. 1994. Design and species selection to reduce urban heat island and conserve energy. In: Proceedings from the 6th national urban forest conference: growing greener communities. Washington, DC: American Forests. 282 p.

Schroeder, H.W.; Cannon, W.N. 1983. The esthetic contribution of trees to residential streets in Ohio towns. Journal of Arboriculture. 9: 237–243.

Schroeder, T. 1982. The relationship of local park and recreation services to residential property values. Journal of Leisure Research. 14: 223–234.

Scott, K.I.; McPherson, E.G.; Simpson, J.R. 1998. Air pollutant uptake by Sacramento's urban forest. Journal of Arboriculture. 24(4): 224–234.

Scott, K.I.; Simpson, J.R.; McPherson, E.G. 1999. Effects of tree cover on parking lot microclimate and vehicle emissions. Journal of Arboriculture. 25(3): 129–142.

Smith, CW. 1998. Impact of alien plants on Hawaii's native biota. http://www.botany.hawaii.edu/faculty/cw_smith/impact.htm. (29 July 2007).

Smith, W.H. 1990. Air pollution and forests. New York: Springer-Verlag. 618 p.

Smith, W.H.; Dochinger, L.S. 1976. Capability of metropolitan trees to reduce atmospheric contaminants. In: Santamour, F.S.; Gerhold, H.D.; Little, S., eds. Better trees for metropolitan landscapes. Gen. Tech. Rep. NE-22. Upper Darby [Newtown Square], PA: U.S. Department of Agriculture, Forest Service, Northeastern Research Station: 49–60.

Sullivan, W.C.; Kuo, E.E. 1996. Do trees strengthen urban communities, reduce domestic violence? Arborist News. 5(2): 33–34.

Summit, J.; McPherson, E.G. 1998. Residential tree planting and care: a study of attitudes and behavior in Sacramento, California. Journal of Arboriculture. 24(2): 89–97.

Sydnor, T.D.; Gamstetter, D.; Nichols, J.; Bishop, B.; Favorite, J.; Blazer, C.; Turpin, L. 2000. Trees are not the root of sidewalk problems. Journal of Arboriculture. 26: 20–29.

Taha, H. 1996. Modeling impacts of increased urban vegetation on ozone air quality in the South Coast Air Basin. Atmospheric Environment. 30: 3423–3430.

Thompson, J.; Nowak, D.J.; Crane, D.E.; Hunkins, J.A. 2004. Iowa, U.S., communities benefit from a tree-planting program: characteristics of recently planted trees. Journal of Arboriculture. 30(10): 1–10.

Thompson, R.P.; Ahern, J.J. 2000. The state of urban and community forestry in California. San Luis Obispo, CA: Urban Forest Ecosystems Institute, California Polytechnic State University. 48 p.

Tretheway, R.; Manthe, A. 1999. Skin cancer prevention: another good reason to plant trees. In: McPherson, E.G.; Mathis, S., eds. Proceedings of the Best of the West summit. Davis, CA: University of California: 72–75.

Tschantz, B.A.; Sacamano, P.L. 1994. Municipal tree management in the United States. Kent, OH: Davey Resource Group.

Tyrvainen, L. 1999. Monetary valuation of urban forest amenities in Finland. Res. Pap. 739. Vantaa, Finland: Finnish Forest Research Institute. 129 p.

Ulrich, R.S. 1985. Human responses to vegetation and landscapes. Landscape and Urban Planning. 13: 29–44.

University of Florida Institute of Food and Agricultural Sciences Extension [UF IFAS Extension]. 2006a. Landscape plants: pruning shade trees in the landscape. http://hort.ifas.ufl.edu/woody/pruning/index.htm. (15 October 2007).

University of Florida Institute of Food and Agricultural Sciences Extension [UF IFAS Extension]. 2006b. Producing quality trees. http://hort.ifas.ufl.edu/woody/nursery_production.html. (29 March 2008).

Urban, J. 1992. Bringing order to the technical dysfunction within the urban forest. Journal of Arboriculture. 18(2): 85–90.

Urban Forest Ecosystems Institute. 2008. Standards for purchasing container-grown landscape trees. http://www.ufei.org/Standards&Specs.html. (28 March 2008).

Urban Horticulture Institute. 2003. Recommended urban trees: site assessment and tree selection for stress tolerance. http://www.hort.cornell.edu/UHI/outreach/recurbtree/index.html. (7 February 2007).

U.S. Census Bureau. 2006. State and county quick facts. http://quickfacts.census.gov/qfd/index.html. (March 2006).

U.S. Department of Agriculture, Forest Service [USDA FS]. 2006a. i-Tree: tools for assessing and managing community forests. http://www.itreetools.org. (27 January 2007).

U.S. Department of Agriculture, Forest Service [USDA FS]. 2006b. Urban tree cover and air quality planning. http://www.treescleanair.org. (21 January 2007).

U.S. Department of Agriculture, Natural Resource Conservation Service, Agroforestry Center [USDA NRCS Agroforesty Center]. 2005. Working trees for treating waste. Western Arborist. 31: 50–52.

U.S. Department of Transportation. 1995. Highway traffic noise analysis and abatement policy and guidance. Washington, DC: U.S. Department of Transportation, Federal Highway Administration. 67 p.

U.S. Environmental Protection Agency [US EPA]. 1998. Ap-42 Compilation of air pollutant emission factors. 5th ed. Research Triangle Park, NC. Volume I. [Pages unknown.]

U.S. Environmental Protection Agency [US EPA]. 2000. 2000 National water quality inventory. http://www.epa.gov/305b/2000report/. (14 September 2007).

U.S. Environmental Protection Agency [US EPA]. 2006a. Air quality system reports. http://www.epa.gov/air/data/reports.html. (21 November 2006).

U.S. Environmental Protection Agency [US EPA]. 2006b. E-GRID (E-GRID2006 Edition). http://www.epa.gov/cleanenergy/egrid/index.htm. (6 December 2006).

van Rentergehm, T.; Botteldooren, D.; Cornelis, W.M.; Gabriels, D. 2002. Reducing screen-induced refraction of noise barriers in wind by vegetative screens. Acta Acustica United with Acustica. 88: 231–238.

Vargas, K.E.; McPherson, E.G.; Simpson J.R.; Peper, P.J.; Gardner, S.L.; Xiao, Q. 2007a. City of Honolulu, Hawaii, municipal forest resource analysis. Internal Tech. Rep. Davis, CA: U.S. Department of Agriculture, Forest Service, Pacific Southwest Research Station, Center for Urban Forest Research. 75 p.

Vargas, K.E.; McPherson, E.G.; Simpson J.R.; Peper, P.J.; Gardner, S.L.; Xiao, Q. 2007b. Interior West community tree guide: benefits, costs, and strategic planting. Gen Tech. Rep. PSW-GTR-205. Albany, CA: U.S. Department of Agriculture, Forest Service, Pacific Southwest Research Station. 105 p.

Vargas, K.E.; McPherson, E.G.; Simpson J.R.; Peper, P.J.; Gardner, S.L.; Xiao, Q. 2007c. Temperate Interior West community tree guide: benefits, costs, and strategic planting. Gen Tech. Rep. PSW-GTR-206. Albany, CA: U.S. Department of Agriculture, Forest Service, Pacific Southwest Research Station. 108 p.

Wang, M.Q.; Santini, D.J. 1995. Monetary values of air pollutant emissions in various U.S. regions. Transportation Research Record 1475: 33–41.

Watson, G.W.; Himelick, E.B. 1997. Principles and practice of planting trees and shrubs. Savoy, IL: International Society of Arboriculture. 199 p.

Wolf, K.L., 1999. Nature and commerce: human ecology in business districts. In: Kollin, C., ed. Building cities of green: proceedings of the 1999 national urban forest conference. Washington, DC: American Forests: 56–59.

Xiao, Q.; McPherson, E.G. 2002. Rainfall interception by Santa Monica's municipal urban forest. Urban Ecosystems. 6: 291–302.

Xiao, Q.; McPherson, E.G.; Simpson, J.R.; Ustin, S.L. 1998. Rainfall interception by Sacramento's urban forest. Journal of Arboriculture. 24(4): 235–244.

Xiao, Q.; McPherson, E.G.; Simpson, J.R.; Ustin, S.L. 2000. Winter rainfall interception by two mature open grown trees in Davis, California. Hydrological Processes. 14(4): 763–784.

Appendix 1: Additional Resources

Additional information regarding urban and community forestry program design and implementation can be obtained from the following sources:

Utilizing Municipal Trees: Ideas From Across the Country by S.M. Bratkovich

Urban Forestry: Planning and Managing Urban Greenspaces by R.W. Miller

An Introductory Guide to Community and Urban Forestry in Washington, Oregon, and California by R. Morgan

A Technical Guide to Urban and Community Forestry by R. Morgan

Urban Tree Risk Management: A Community Guide to Program Design and Implementation edited by J.D. Pokorny

For additional information on tree selection, planting, establishment, and care see the following resources:

How to Prune Trees by P.J. Bedker, J.G. O'Brien, and M.E. Mielke

Training Young Trees for Structure and Form, a video by L.R. Costello

An Illustrated Guide to Pruning by E.F. Gilman

Trees for Urban and Suburban Landscapes by E.F. Gilman

Planting Trees and Shrubs for Long-Term Health by R. Hargrave, G.R. Johnson, and M.E. Zins

Arboriculture, 4th ed., by R.W. Harris, J.R. Clark, and N.P. Matheny

Native Trees, Shrubs, and Vines for Urban and Rural America by G.L. Hightshoe

Ornamental Plants for Subtropical Regions by R.S. Hoyt

Street Trees Recommended for Southern California, 2nd ed. by M.T. Mahoney, A.H. Remyn, M.P. Trotter, W.R. Trotter, M.E. Chamness, and K.J. Greby

Principles and Practice of Planting Trees and Shrubs by G.W. Watson and E.B. Himelick

Alliance for Community Trees: http://actrees.org

International Society of Arboriculture: http://www.isa-arbor.com, including their *Tree City USA Bulletin* series by J.R. Fazio

National Arbor Day Foundation: http://www.arborday.org

TreeLink: http://www.treelink.org

The Urban Horticulture Institute: http://www.hort.cornell.edu/UHI/outreach/recurbtree/index.html

These suggested references are only a starting point. Your local cooperative extension agent, urban forester, or state forestry agency can provide you with up-to-date and local information.

Appendix 2: Benefit-Cost Information Tables

Information in this appendix can be used to estimate benefits and costs associated with proposed tree plantings. The tables contain data for representative small (silver buttonwood tree), medium (rainbow shower tree), and large (monkeypod) trees (see "Common and Scientific Names" section). Data are presented as annual values for each 5-year interval after planting (tables 6 to 14). Annual values incorporate effects of tree loss. Based on the results of our survey, we assume that 19 percent of the trees planted die by the end of the 40-year period.

For the benefits tables (tables 6, 9, 12), there are two columns for each 5-year interval. In the first column, values describe **resource units** (RUs): for example, the amount of air conditioning energy saved in kilowatt hours per year per tree, air pollutant uptake in pounds per year per tree, and rainfall intercepted in gallons per year per tree. Energy and carbon dioxide (CO_2) benefits for residential yard trees are broken out by tree location to show how shading effects differ among trees opposite west-, south-, and east-facing building walls. The second column for each 5-year interval contains dollar values obtained by multiplying RUs by local prices (e.g., kWh saved [RU] x $/kWh).

In the costs tables (tables 7, 10, 13), costs are broken down into categories for yard and public trees. Costs for yard trees do not differ by planting location (i.e., east, west, south walls). Although tree purchase and planting costs occur at year 1, we divided this value by 5 years to derive an average annual cost for the first 5-year period. All other costs are the estimated values for each year and not values averaged over 5 years.

Annual net benefits are calculated by subtracting annual costs from annual benefits and are presented in tables 8, 11, and 14. Data are presented for a yard tree opposite west-, south-, and east-facing walls, as well as for the public tree.

The last column in each table presents 40-year-average annual values. These numbers were calculated by dividing the total costs and benefits by 40 years.

Table 6—Annual benefits at 5-year intervals and 40-year average for a representative small tree (silver buttonwood)

	Year 5 RU	Year 5 Value $	Year 10 RU	Year 10 Value $	Year 15 RU	Year 15 Value $	Year 20 RU	Year 20 Value $	Year 25 RU	Year 25 Value $	Year 30 RU	Year 30 Value $	Year 35 RU	Year 35 Value $	Year 40 RU	Year 40 Value $	40-year average RU	40-year average Value $
Cooling (kWh):																		
Yard: west	3	0.37	17	2.01	43	5.25	73	8.95	126	15.37	178	21.77	245	29.88	300	36.54	123	15.02
Yard: south	1	0.14	9	1.05	22	2.63	36	4.42	65	7.98	95	11.61	134	16.38	173	21.10	67	8.17
Yard: east	2	0.30	14	1.66	35	4.22	58	7.13	102	12.44	145	17.73	200	24.43	247	30.12	100	12.25
Public	1	0.14	9	1.05	18	2.24	28	3.41	40	4.93	52	6.35	65	7.98	79	9.60	37	4.46
Net carbon dioxide (lbs):																		
Yard: west	8	0.03	40	0.14	104	0.35	179	0.60	316	1.06	459	1.53	646	2.16	841	2.81	324	1.08
Yard: south	5	0.02	27	0.09	67	0.22	115	0.38	211	0.71	314	1.05	454	1.52	621	2.08	227	0.76
Yard: east	7	0.02	36	0.12	90	0.30	153	0.51	275	0.92	401	1.34	568	1.90	750	2.50	285	0.95
Public	5	0.02	27	0.09	62	0.21	100	0.34	168	0.56	239	0.80	335	1.12	458	1.53	174	0.58
Air pollution (lb):[a]																		
Ozone uptake	0.007	0.01	0.034	0.05	0.075	0.11	0.118	0.17	0.175	0.26	0.231	0.34	0.298	0.44	0.361	0.53	0.162	0.24
Nitrogen dioxide uptake + avoided	0.011	0.02	0.066	0.10	0.163	0.24	0.271	0.40	0.457	0.67	0.643	0.95	0.878	1.29	1.085	1.60	0.447	0.66
Sulfur dioxide uptake + avoided	0.010	0.01	0.058	0.09	0.143	0.22	0.237	0.36	0.401	0.61	0.564	0.86	0.770	1.17	0.952	1.45	0.392	0.60
Small particulate matter uptake + avoided	0.006	0.01	0.039	0.05	0.102	0.14	0.182	0.24	0.274	0.37	0.363	0.49	0.457	0.61	0.544	0.73	0.246	0.33
Volatile organic compounds avoided	0.002	0.00	0.011	0.01	0.027	0.02	0.045	0.03	0.077	0.05	0.108	0.07	0.148	0.09	0.183	0.11	0.075	0.05
Biogenic volatile organic compounds released	-0.001	0.00	-0.002	0.00	-0.005	0.00	-0.009	-0.01	-0.014	-0.01	-0.019	-0.01	-0.026	-0.02	-0.033	-0.02	-0.014	-0.01
Pollution avoided + net uptake	0.034	0.05	0.206	0.29	0.504	0.72	0.844	1.20	1.370	1.95	1.890	2.69	2.525	3.59	3.094	4.40	1.308	1.86
Hydrology (gal):[a]																		
Rainfall interception	37	0.37	134	1.34	276	2.76	425	4.25	631	6.31	842	8.42	1108	11.08	1387	13.87	605	6.05
Aesthetics and other:																		
Yard		7.73		9.72		12.17		14.98		18.20		21.85		25.96		30.54		17.64
Public		8.75		11.01		13.78		16.97		20.61		24.75		29.40		34.59		19.98
Total benefits:																		
Yard: west		8.54		13.51		21.23		29.97		42.88		56.26		72.67		88.15		41.65
Yard: south		8.30		12.50		18.49		25.23		35.14		45.62		58.53		71.98		34.48
Yard: east		8.47		13.14		20.16		28.07		39.81		52.03		66.95		81.43		38.76
Public		9.33		13.79		19.70		26.16		34.36		43.00		53.17		63.98		32.94

Note: Annual values incorporate effects of tree loss. We assume that 5 percent of trees planted die during the first 5 years and 14 percent during the remaining 35 years for a total mortality of 19 percent. RU = resource units.

[a] Values are the same for yard and public trees.

Table 7—Annual costs (dollars per tree) at 5-year intervals and 40-year average for a representative small tree (silver buttonwood)

Costs	Year 5	Year 10	Year 15	Year 20	Year 25	Year 30	Year 35	Year 40	40-year average
					Dollars				
Tree and planting:[a]									
Yard	60								7.5
Public	24								3
Pruning:									
Yard	0.23	0.15	0.14	0.14	0.14	5.01	4.88	4.75	1.58
Public	9.50	6.13	5.99	5.85	5.70	18.08	17.62	17.16	9.92
Remove and dispose:									
Yard	0.20	0.60	0.97	1.40	1.89	2.43	3.03	3.68	1.59
Public	0.20	0.38	0.62	0.89	1.20	1.55	1.93	2.34	1.02
Pests and disease:									
Yard	0	0	0	0	0	0	0	0	0
Public	0.06	0.12	0.19	0.26	0.35	0.43	0.53	0.62	0.29
Infrastructure repair:									
Yard	0.13	0.28	0.45	0.63	0.83	1.04	1.27	1.50	0.69
Public	1.07	2.25	3.58	5.05	6.64	8.34	10.13	11.99	5.53
Cleanup:									
Yard	0.04	0.09	0.15	0.21	0.28	0.35	0.42	0.50	0.23
Public	0.36	0.75	1.19	1.68	2.21	2.78	3.38	4.00	1.84
Admin./inspect/other:									
Yard	0	0	0	0	0	0	0	0	0
Public	0.40	0.84	1.34	1.89	2.49	3.13	3.80	4.50	2.07
Total costs:									
Yard	60.61	1.12	1.71	2.38	3.13	8.82	9.59	10.43	11.58
Public	35.58	10.47	12.91	15.63	18.60	34.31	37.37	40.62	23.67

Note: Annual values incorporate effects of tree loss. We assume that 5 percent of trees planted die during the first 5 years and 14 percent during the remaining 35 years for a total mortality of 19 percent.

[a] Although tree and planting costs occur in year 1, this value was divided by 5 years to derive an average annual cost for the first 5-year period.

Table 8—Annual net benefits (dollars per tree) at 5-year intervals and 40-year average for a representative small tree (silver buttonwood)

Total net benefits	Year 5	Year 10	Year 15	Year 20	Year 25	Year 30	Year 35	Year 40	40-year average
					Dollars				
Yard: west	-52	12	20	28	40	47	63	78	30
Yard: south	-52	11	17	23	32	37	49	62	23
Yard: east	-52	12	18	26	37	43	57	71	27
Public	-26	3	7	11	16	9	16	23	9

Note: Annual values incorporate effects of tree loss. We assume that 5 percent of trees planted die during the first 5 years and 14 percent during the remaining 35 years for a total mortality of 19 percent. See table 6 for annual benefits and table 7 for annual costs.

Table 9—Annual benefits (dollars per tree) at 5-year intervals and 40-year average for a representative medium tree (rainbow shower tree)

	Year 5		Year 10		Year 15		Year 20		Year 25		Year 30		Year 35		Year 40		40-year average	
	RU	Value	RU	Value	RU	Value	RU	Value	RU	Value	RU	Value	RU	Value	RU	Value	RU	Value
		$		$		$		$		$		$		$		$		$
Cooling (kWh):																		
Yard: west	33	3.99	105	12.82	194	23.64	273	33.26	336	41.02	397	48.39	435	53.05	470	57.36	280	34.19
Yard: south	15	1.83	50	6.12	100	12.20	152	18.53	200	24.44	246	30.06	288	35.17	328	40.03	173	21.05
Yard: east	25	3.07	83	10.07	156	18.99	222	27.09	277	33.75	329	40.09	364	44.40	397	48.41	231	28.23
Public	11	1.35	30	3.71	51	6.19	71	8.69	90	10.96	108	13.12	123	14.98	137	16.74	78	9.47
Net carbon dioxide (lbs):																		
Yard: west	67	0.22	208	0.69	376	1.25	525	1.75	646	2.16	761	2.54	833	2.78	900	3.01	540	1.80
Yard: south	36	0.12	112	0.38	213	0.71	316	1.06	411	1.37	501	1.67	580	1.94	654	2.18	353	1.18
Yard: east	54	0.18	168	0.56	310	1.03	438	1.46	543	1.81	644	2.15	711	2.37	773	2.58	455	1.52
Public	29	0.10	78	0.26	128	0.43	176	0.59	219	0.73	260	0.87	293	0.98	323	1.08	188	0.63
Air pollution (lb):[a]																		
Ozone uptake	0.039	0.06	0.111	0.16	0.191	0.28	0.273	0.40	0.349	0.51	0.425	0.63	0.492	0.72	0.558	0.82	0.305	0.45
Nitrogen dioxide uptake + avoided	0.114	0.17	0.363	0.53	0.675	0.99	0.969	1.43	1.220	1.80	1.458	2.15	1.637	2.41	1.805	2.66	1.030	1.52
Sulfur dioxide uptake + avoided	0.100	0.15	0.319	0.49	0.593	0.90	0.851	1.30	1.071	1.63	1.281	1.95	1.438	2.19	1.586	2.42	0.905	1.38
Small particulate matter uptake + avoided	0.027	0.04	0.117	0.16	0.264	0.35	0.442	0.59	0.609	0.82	0.768	1.03	0.912	1.22	0.940	1.26	0.510	0.68
Volatile organic compounds avoided	0.019	0.01	0.062	0.04	0.115	0.07	0.165	0.10	0.207	0.12	0.247	0.15	0.277	0.17	0.306	0.18	0.175	0.11
Biogenic volatile organic compounds released	-0.001	0.00	-0.004	0.00	-0.008	0.00	-0.015	-0.01	-0.023	-0.01	-0.032	-0.02	-0.041	-0.02	-0.041	-0.02	-0.021	-0.01
Pollution avoided + net uptake	0.299	0.43	0.968	1.38	1.830	2.60	2.685	3.81	3.434	4.87	4.149	5.88	4.715	6.69	5.153	7.31	2.904	4.12
Hydrology (gal):[a]																		
Rainfall interception	185	1.85	474	4.74	781	7.81	1100	11.00	1401	14.01	1702	17.02	1986	19.86	2270	22.70	1237	12.37
Aesthetics and other:																		
Yard		39.74		40.10		40.62		40.71		40.43		39.79		38.85		37.64		39.74
Public		45.01		45.42		46.01		46.11		45.79		45.07		44.01		42.63		45.01
Total benefits:																		
Yard: west		46.22		59.73		75.92		90.54		102.48		113.62		121.23		128.03		92.22
Yard: south		43.96		52.71		63.94		75.10		85.11		94.43		102.51		109.87		78.45
Yard: east		45.26		56.85		71.06		84.08		94.87		104.93		112.17		118.65		85.98
Public		48.73		55.51		63.04		70.20		76.36		81.96		86.51		90.47		71.60

Note: Annual values incorporate effects of tree loss. We assume that 5 percent of trees planted die during the first 5 years and 14 percent during the remaining 35 years for a total mortality of 19 percent. RU = resource units.

[a] Values are the same for yard and public trees.

Table 10—Annual costs (dollars per tree) at 5-year intervals and 40-year average for a representative medium tree (rainbow shower tree)

Costs	Year 5	Year 10	Year 15	Year 20	Year 25	Year 30	Year 35	Year 40	40-year average
					Dollars				
Tree and planting:[a]									
Yard	60								7.5
Public	24								3
Pruning:									
Yard	0.23	0.15	5.39	5.26	5.13	5.01	4.88	4.75	3.89
Public	9.50	6.13	19.46	19.00	18.54	18.08	17.62	17.16	15.86
Remove and dispose:									
Yard	0.20	0.75	1.11	1.46	1.79	2.11	2.41	2.70	1.45
Public	0.20	0.48	0.71	0.93	1.14	1.34	1.53	1.72	0.93
Pests and disease:									
Yard	0	0	0	0	0	0	0	0	0
Public	0.08	0.15	0.21	0.27	0.33	0.38	0.42	0.46	0.26
Infrastructure repair:									
Yard	0.18	0.35	0.51	0.66	0.79	0.90	0.84	1.10	0.63
Public	1.44	2.84	4.11	5.27	6.31	7.24	8.06	8.79	5.07
Cleanup:									
Yard	0.06	0.12	0.17	0.22	0.26	0.30	0.28	0.37	0.21
Public	0.48	0.95	1.37	1.76	2.10	2.41	2.69	2.93	1.69
Admin./inspect/other:									
Yard	0	0	0	0	0	0	0	0	0
Public	0.54	1.06	1.54	1.97	2.37	2.71	3.02	3.30	1.90
Total costs									
Yard	60.67	1.37	7.19	7.60	7.98	8.32	8.40	8.91	13.68
Public	36.24	11.60	27.40	29.20	30.78	32.16	33.35	34.35	28.72

Note: Annual values incorporate effects of tree loss. We assume that 5 percent of trees planted die during the first 5 years and 14 percent during the remaining 35 years for a total mortality of 19 percent.

[a]Although tree and planting costs occur in year 1, this value was divided by 5 years to derive an average annual cost for the first 5-year period.

Table 11—Annual net benefits (dollars per tree) at 5-year intervals and 40-year average for a representative medium tree (rainbow shower tree)

Total net benefits	Year 5	Year 10	Year 15	Year 20	Year 25	Year 30	Year 35	Year 40	40-year average
					Dollars				
Yard: west	-14	58	69	83	95	105	113	119	79
Yard: south	-17	51	57	68	77	86	94	101	65
Yard: east	-15	55	64	76	87	97	104	110	72
Public	12	44	36	41	46	50	53	56	43

Note: Annual values incorporate effects of tree loss. We assume that 5 percent of trees planted die during the first 5 years and 14 percent during the remaining 35 years for a total mortality of 19 percent. See table 9 for annual benefits and table 10 for annual costs.

Table 12—Annual benefits (dollars per tree) at 5-year intervals and 40-year average for a representative large tree (monkeypod)

	Year 5		Year 10		Year 15		Year 20		Year 25		Year 30		Year 35		Year 40		40-year average	
	RU	Value	RU	Value	RU	Value	RU	Value	RU	Value	RU	Value	RU	Value	RU	Value	RU	Value
		$		$		$		$		$		$		$		$		$
Cooling (kWh):																		
Yard: west	24	2.90	73	8.91	138	16.88	207	25.23	277	33.82	344	41.97	405	49.43	464	56.65	242	29.48
Yard: south	17	2.02	54	6.60	109	13.32	168	20.54	243	29.60	314	38.30	380	46.41	445	54.25	216	26.38
Yard: east	21	2.60	66	8.06	126	15.42	190	23.17	259	31.58	324	39.58	384	46.89	443	53.99	227	27.66
Public	15	1.84	44	5.37	84	10.27	127	15.46	179	21.79	229	28.00	282	34.41	334	40.77	162	19.74
Net carbon dioxide (lbs):																		
Yard: west	56	0.19	158	0.53	291	0.97	431	1.44	578	1.93	719	2.40	852	2.85	982	3.28	508	1.70
Yard: south	43	0.15	125	0.42	241	0.80	364	1.22	518	1.73	667	2.23	809	2.70	947	3.16	464	1.55
Yard: east	52	0.17	146	0.49	270	0.90	402	1.34	546	1.82	685	2.29	816	2.73	944	3.15	483	1.61
Public	41	0.14	108	0.36	197	0.66	292	0.98	407	1.36	520	1.74	639	2.13	756	2.52	370	1.24
Air pollution (lb):[a]																		
Ozone uptake	0.050	0.07	0.157	0.23	0.301	0.44	0.459	0.68	0.650	0.96	0.846	1.25	1.060	1.56	1.278	1.88	0.600	0.88
Nitrogen dioxide uptake + avoided	0.106	0.16	0.330	0.49	0.636	0.94	0.962	1.42	1.333	1.96	1.690	2.49	2.031	2.99	2.364	3.48	1.182	1.74
Sulfur dioxide uptake + avoided	0.093	0.14	0.288	0.44	0.557	0.85	0.841	1.28	1.166	1.78	1.478	2.25	1.776	2.71	2.068	3.15	1.033	1.58
Small particulate matter uptake + avoided	0.019	0.03	0.083	0.11	0.231	0.31	0.477	0.64	0.772	1.03	1.089	1.46	1.421	1.90	1.762	2.36	0.732	0.98
Volatile organic compounds avoided	0.018	0.01	0.054	0.03	0.105	0.06	0.159	0.10	0.220	0.13	0.278	0.17	0.333	0.20	0.387	0.23	0.194	0.12
Biogenic volatile organic compounds released	0.000	0.00	-0.002	0.00	-0.005	0.00	-0.010	-0.01	-0.018	-0.01	-0.027	-0.02	-0.039	-0.02	-0.052	-0.03	-0.019	-0.01
Pollution avoided + net uptake	0.286	0.41	0.911	1.30	1.825	2.60	2.887	4.10	4.122	5.85	5.354	7.60	6.582	9.34	7.807	11.08	3.722	5.28
Hydrology (gal):[a]																		
Rainfall interception	218	2.18	576	5.76	1058	10.58	1585	15.85	2247	22.47	2934	29.34	3713	37.13	4529	45.29	2108	21.08
Aesthetics and other:																		
Yard		24.77		32.59		40.43		47.89		54.99		61.71		68.06		74.04		50.56
Public		28.05		36.92		45.79		54.25		62.28		69.90		77.09		83.86		57.27
Total benefits:																		
Yard: west		30.45		49.09		71.46		94.52		119.06		143.01		166.81		190.34		108.09
Yard: south		29.53		46.67		67.74		89.61		114.64		139.17		163.64		187.83		104.85
Yard: east		30.14		48.20		69.93		92.36		116.71		140.51		164.14		187.56		106.19
Public		32.62		49.70		69.90		90.64		113.75		136.56		160.10		183.53		104.60

Note: Annual values incorporate effects of tree loss. We assume that 5 percent of trees planted die during the first 5 years and 14 percent during the remaining 35 years for a total mortality of 19 percent. RU = resource unit.

[a] Values are the same for yard and public trees.

Table 13—Annual costs (dollars per tree) at 5-year intervals and 40-year average for a representative large tree (monkeypod)

Costs	Year 5	Year 10	Year 15	Year 20	Year 25	Year 30	Year 35	Year 40	40-year average
					Dollars				
Tree and planting:[a]									
Yard	60								7.5
Public	24								3
Pruning:									
Yard	0.23	0.15	5.39	5.26	5.13	5.01	4.88	15.84	4.72
Public	6.27	19.92	19.46	19.00	18.54	18.08	17.62	25.00	16.91
Remove and dispose:									
Yard	1.57	1.15	1.62	2.10	2.58	3.05	3.53	4.00	2.20
Public	1.00	0.73	1.03	1.34	1.64	1.94	2.24	2.55	1.40
Pests and disease:									
Yard	0	0	0	0	0	0	0	0	0
Public	0.14	0.23	0.31	0.39	0.47	0.55	0.61	0.68	0.39
Infrastructure repair:									
Yard	0.33	0.54	0.75	0.95	1.13	1.31	1.48	1.63	0.94
Public	2.61	4.35	6.00	7.58	9.07	10.48	11.81	13.05	7.51
Cleanup:									
Yard	0.11	0.18	0.25	0.32	0.38	0.44	0.49	0.54	0.31
Public	0.87	1.45	2.00	2.53	3.02	3.49	3.94	4.35	2.50
Admin./inspect/other:									
Yard	0	0	0	0	0	0	0	0	0
Public	0.98	1.63	2.25	2.84	3.40	3.93	4.43	4.89	2.82
Total costs:									
Yard	62.23	2.02	8.01	8.62	9.22	9.80	10.37	22.02	15.68
Public	35.86	28.30	31.06	33.68	36.15	38.47	40.65	50.52	34.54

Note: Annual values incorporate effects of tree loss. We assume that 5 percent of trees planted die during the first 5 years and 14 percent during the remaining 35 years for a total mortality of 19 percent.

[a]Although tree and planting costs occur in year 1, this value was divided by 5 years to derive an average annual cost for the first 5-year period.

Table 14—Annual net benefits (dollars per tree) at 5-year intervals and 40-year average for a representative large tree (monkeypod)

Total net benefits	Year 5	Year 10	Year 15	Year 20	Year 25	Year 30	Year 35	Year 40	40-year average
					Dollars				
Yard: west	-32	47	63	86	110	133	156	168	92
Yard: south	-33	45	60	81	105	129	153	166	89
Yard: east	-32	46	62	84	107	131	154	166	91
Public	-3	21	39	57	78	98	119	133	70

Note: Annual values incorporate effects of tree loss. We assume that 5 percent of trees planted die during the first 5 years and 14 percent during the remaining 35 years for a total mortality of 19 percent. See table 12 for annual benefits and table 13 for annual costs.

Appendix 3: Procedures for Estimating Benefits and Costs

Approach

Pricing Benefits and Costs

In this study, annual benefits and costs over a 40-year planning horizon were estimated for newly planted trees in three residential yard locations (east, south, and west of the dwelling unit) and a public streetside or park location. Trees in these hypothetical locations are called "yard" and "public" trees, respectively. Prices were assigned to each cost (e.g., planting, pruning, removal, irrigation, infrastructure repair, liability) and benefit (e.g., heating/cooling energy savings, air-pollution reduction, stormwater-runoff reduction) through direct estimation and implied valuation of benefits as environmental externalities. This approach made it possible to estimate the net benefits of plantings in "typical" locations with "typical" tree species.

To account for differences in the mature size and growth rates of different tree species, we report results for a small (silver buttonwood tree), medium (rainbow shower tree), and large (monkeypod) tree (see "Common and Scientific Names" section). Results are reported for 5-year intervals for 40 years.

Mature tree height is frequently used to characterize small, medium, and large species because matching tree height to available overhead space is an important design consideration. However, in this analysis, leaf surface area (LSA) and crown diameter were also used to characterize **mature tree size**. These additional measurements are useful indicators for many functional benefits of trees that relate to leaf-atmosphere processes (e.g., interception, transpiration, photosynthesis). Tree growth rates, dimensions, and LSA estimates are based on tree growth modeling.

Growth Modeling

Growth models are based on data collected in Honolulu, Hawaii. Using Honolulu's partial street tree inventory that included 43,817 trees, we measured a stratified random sample of 20 of the most common tree species to establish relations between tree age, size, leaf area, and biomass. The species were as follows:

- Hong Kong orchid tree (*Bauhinia* x *blakeana* Dunn)
- Ironwood (*Casuarina equisetifolia* L.)
- Kamani (*Calophyllum inophyllum* L.)

- Rainbow shower tree (*Cassia* x *nealiae* H.S. Irwin & Barneby)
- Fiddlewood (*Citharexylum spinosum* L.)
- Silver buttonwood (*Conocarpus erectus* L. var. *argenteus* Millsp.)
- Kou (*Cordia subcordata* Lam.)
- Royal poinciana (*Delonix regia* (Bojer ex Hook.) Raf.)
- False olive (*Elaeodendron orientale* J. Jacq.)
- Benjamin fig (*Ficus benjamina* L.)
- Fern tree (*Filicium decipiens* (Wight & Arn.) Thwaites ex Hook. f.)
- Paraguay-tea (*Ilex paraguariensis* St.-Hil.)
- Giant crapemyrtle (*Lagerstroemia speciosa* (L.) Pers.)
- Paperbark (*Melaleuca quinquenervia* (Cav.) Blake)
- Monkeypod (*Samanea saman* (Jacq.) Merr.)
- West Indian mahogany (*Swietenia mahagoni* (L.) Jacq.)
- Silver trumpet tree (*Tabebuia aurea* (Manso) Benth. & Hook. f. ex S. Moore)
- Golden trumpet tree (*Tabebuia ochracea* subsp. *neochrysantha* (A.H. Gentry) A.H. Gentry)
- Pink tecoma (*Tabebuia heterophylla* (DC.) Britt)
- Coconut palm (*Cocos nucifera* L.)
- Date palm (*Phoenix dactylifera* L.)
- Manila palm (*Veitchia merrillii* (Becc.) H.E. Moore)

For the growth models, information spanning the life cycle of predominant tree species was collected. The inventory was stratified into the following nine diameter-at-breast-height (d.b.h.) classes:

- 0 to 2.9 in
- 3.0 to 5.9 in
- 6.0 to 11.9 in
- 12.0 to 17.9 in
- 18.0 to 23.9 in
- 24.0 to 29.9 in
- 30.0 to 35.9 in
- 36.0 to 41.9 in
- >42.0 in

Thirty-five to sixty trees of each species were randomly selected for surveying, along with an equal number of alternative trees. Tree measurements included d.b.h. (to nearest 0.1 cm [0.04 in] by sonar measuring device), tree crown and bole height (to nearest 0.5 m [1.6 ft] by clinometer), crown diameter in two directions (parallel and perpendicular to nearest street to nearest 0.5 m [1.6 ft] by sonar measuring device), tree condition, and location. Replacement trees were sampled when trees from the original sample population could not be located. Tree age was determined by street-tree managers. Field work was conducted in September 2005.

Crown volume and leaf area were estimated from computer processing of tree-crown images obtained with a digital camera. The method has shown greater accuracy than other techniques (±20 percent of actual leaf area) in estimating crown volume and leaf area of open-grown trees (Peper and McPherson 2003).

Linear regression was used to fit predictive models with d.b.h. as a function of age for each of the 20 sampled species. Predictions of LSA, crown diameter, and height metrics were modeled as a function of d.b.h. by using best-fit models. After inspecting the growth curves for each species, we selected the typical small, medium, and large tree species for this report.

Reporting Results

Results are reported in terms of annual values per tree planted. However, to make these calculations realistic, mortality rates are included. Based on our survey of regional municipal foresters and commercial arborists, this analysis assumed that 19 percent of the hypothetical planted trees died over the 40-year period. Annual mortality rates were 1.0 percent for the first 5 years, and 0.40 percent per year after that. The accounting approach "grows" trees in different locations and uses computer simulation to directly calculate the annual flow of benefits and costs as trees mature and die (McPherson 1992).

Benefits and costs are directly connected with tree-size variables such as trunk d.b.h., tree canopy cover, and LSA. For instance, pruning and removal costs usually increase with tree size, expressed as d.b.h. For some parameters, such as sidewalk repair, costs are negligible for young trees but increase relatively rapidly as tree roots grow large enough to heave pavement. For other parameters, such as air-pollutant uptake and rainfall interception, benefits are related to tree canopy cover and leaf area.

Most benefits occur on an annual basis, but some costs are periodic. For instance, street trees may be pruned on regular cycles but are removed in a less regular fashion (e.g., when they pose a hazard or soon after they die). In this analysis, most costs and benefits are reported for the year in which they occur. However, periodic costs such as pruning, pest and disease control, and infrastructure repair are presented on an average annual basis. Although spreading one-time costs over each year of a maintenance cycle does not alter the 40-year nominal expenditure, it can lead to inaccuracies if future costs are discounted to the present.

Benefit and Cost Valuation

Source of cost estimates—

Frequency and costs of public tree management were estimated based on surveys with municipal foresters in Honolulu, Hawaii, and Naples and Lantana, Florida. Commercial arborists from Hollywood, Florida, and Ewa Beach, Honolulu, Kaneohe, and Waimanalo, Hawaii, provided information on tree management costs on residential properties.

Monetizing benefits—

To monetize effects of trees on energy use, we take the perspective of a residential customer by using retail electricity and natural gas prices for utilities serving the Hawaiian Islands. The retail price of energy reflects a full accounting of costs as paid by the end user, such as the utility costs of power generation, transmission, distribution, administration, marketing, and profit. This perspective aligns with our modeling method, which calculates energy effects of trees based on differences among consumers in heating and air conditioning equipment types, saturations, building construction types, and base loads.

The preferred way to value air quality benefits from trees is to first determine the costs of damages to human health from polluted air, then calculate the value of avoided costs because trees are cleaning the air. Economic valuation of damages to human health usually uses information on willingness to pay to avoid damages obtained via interviews or direct estimates of the monetary costs of damages (e.g., alleviating headaches, extending life). Empirical correlations developed by Wang and Santini (1995) reviewed five studies and 15 sets of regional cost data to relate per-ton costs of various pollutant emissions to regional ambient air quality measurements and population size. We use their damage-based estimates unless the values are negative, in which case we use their control-cost-based estimates.

Calculating Benefits
Calculating Energy Benefits

The prototypical building used as a basis for the simulations was typical of post-1980 construction practices and represents approximately one-third of the total single-family residential housing stock in the Tropical region. The house was a one-story, wood-frame, slab-on-grade building with a conditioned floor area of 2,070 ft^2, window area (double-glazed) of 263 ft^2, and wall and ceiling insulation of R11 and R25, respectively. The central cooling system had a **seasonal energy efficiency ratio** (SEER) of 10. Building footprints were square, reflecting average impacts for a large number of buildings (McPherson and Simpson 1999). Buildings were simulated with 1.5-ft overhangs. Blinds had a visual density of 37 percent and were assumed to be closed when the air conditioner was operating. Thermostat settings were 78 °F. Because the prototype building was larger, but more energy efficient, than most other construction types, our projected energy savings can be considered similar to those for older, less thermally efficient, but smaller buildings. The energy simulations relied on typical meteorological data from Honolulu (Marion and Urban 1995).

Calculating energy savings—

The dollar value of energy savings was based on regional average residential electricity prices of $0.122/kWh. Homes were assumed to have central air conditioning. Because of the very mild (or nonexistent) winters in the Tropical region, effects of trees on heating costs were negligible and were therefore not considered.

Calculating shade effects—

Residential yard trees were within 60 ft of homes so as to directly shade walls and windows. Shade effects of these trees on building energy use were simulated for small, medium, and large trees at three tree-to-building distances, following methods outlined by McPherson and Simpson (1999). Results of shade effects for each tree were averaged over distance and weighted by occurrence within each of three distance classes: 28 percent at 10 to 20 ft (3 to 6 m), 68 percent at 20 to 40 ft (6 to 12 m), and 4 percent at 40 to 60 ft (12 to 18 m) (McPherson and Simpson 1999).

The small (silver buttonwood) and medium (rainbow shower tree) trees had visual densities of 77 and 64 percent, respectively; the large tree (monkeypod) had visual densities of 77 percent during leaf-on season and 20 percent during leaf-off season.

Foliation periods for deciduous trees were obtained from the literature (Hammond et al. 1980, McPherson 1984) and adjusted for Honolulu based on consultation with the city forester and a local horticulturist from the Honolulu Botanical Gardens (Oka 2006, Sand 2006). The monkeypod was considered semi-deciduous with a leaf-off period of March 1 through 31.

Results are reported for trees shading east-, south-, and west-facing surfaces. Our results for public trees are conservative in that we assumed that they do not provide shading benefits. For example, in Modesto, California, 15 percent of total annual dollar energy savings from street trees was due to shade and 85 percent due to **climate effects** (McPherson et al. 1999a).

Calculating climate effects—

In addition to localized shade effects, which were assumed to accrue only to residential yard trees, lowered air temperatures and windspeeds from increased neighborhood tree cover (referred to as climate effects) produced a net decrease in demand for summer cooling (reduced windspeeds by themselves may increase or decrease cooling demand, depending on the circumstances). Climate effects on energy use, air temperature, and windspeed, as a function of neighborhood canopy cover, were estimated from published values (McPherson and Simpson 1999). Existing tree canopy plus building cover was 20 percent based on estimates of urban tree cover from Elliott (2004). Canopy cover was calculated to increase by 4.3 percent, 6.6 percent, and 8.8 percent for 20-year-old small, medium, and large trees, respectively, based on an effective lot size (actual lot size plus a portion of adjacent street and other rights-of-way) of 10,000 ft^2, and one tree on average was assumed per lot. Climate effects were estimated by simulating effects of wind reductions and air-temperature reductions on energy use. Climate effects accrued for both public and yard trees.

Atmospheric Carbon Dioxide Reduction

Calculating reduction in carbon dioxide emissions from powerplants—

Conserving energy in buildings can reduce carbon dioxide (CO_2) emissions from powerplants. Emission reductions were calculated as the product of energy savings for heating and cooling with CO_2 **emission factors** (table 15) based on data for the Hawaiian islands, where the average fuel mix consists mainly of oil and coal (76 and 15 percent, respectively) (US EPA 2006b). The value of $6.68 per ton CO_2 reduction (table 15) was based on the average value reported by Pearce (2003).

Calculating carbon storage—

Sequestration, the net rate of CO_2 storage in above- and belowground biomass over the course of one growing season, was calculated from tree height and d.b.h. data for species other than conifer and palm using a general biomass equation for tropical dry forest stands (Chave et al. 2005). The dry forest equation was selected based on Honolulu average annual rainfall patterns. Palm biomass was modeled using equations from Frangi and Lugo (1985). For conifers, volume equations

Table 15—Emissions factors and implied values for carbon dioxide and criteria air pollutants

	Emission factor[a]	Implied value[b]
	lb/MWh	*$/lb*
Carbon dioxide	1,733,000	0.003
Nitrogen dioxide	5.206	1.47
Sulfur dioxide	4.566	1.52
Small particulate matter	0.921	1.34
Volatile organic compounds	0.917	0.60

[a] US EPA (2006a), except Ottinger et al. 1990 for volatile organic compounds.

[b] Carbon dioxide from Pearce (2003). Value for others based on the methods of Wang and Santini (1995) using emissions concentrations from US EPA (2006b) and population estimates from the U.S. Census Bureau (2006).

derived from California open-grown trees were used (Pillsbury et al. 1998). Volume estimates were converted to green- and dry-weight estimates (Markwardt 1930) and divided by 78 percent to incorporate root biomass. Dry-weight biomass was converted to carbon (50 percent) and these values were converted to CO_2. The amount of CO_2 sequestered each year is the annual increment of CO_2 stored as trees increase their biomass.

Calculating CO_2 released by power equipment—
Tree-related emissions of CO_2, based on gasoline and diesel fuel consumption during tree care in our survey cities, were calculated by using the value 0.30 lbs of CO_2 per in d.b.h. (Koike 2006). This amount may overestimate CO_2 release associated with less intensively maintained residential yard trees.

Calculating CO_2 released during decomposition—
To calculate CO_2 released through decomposition of dead woody biomass, we conservatively estimated that dead trees were removed and mulched in the year that death occurred, and that 80 percent of their stored carbon was released to the atmosphere as CO_2 in the same year (McPherson and Simpson 1999).

Calculating Reduction in Air Pollutant Emissions

Reductions in building energy use also result in reduced emission of air pollutants from powerplants and space-heating equipment. Volatile organic hydrocarbons (VOCs) and nitrogen dioxide (NO_2)—both precursors of ozone (O_3) formation—as well as sulfur dioxide (SO_2) and particulate matter of <10 micron diameter (PM_{10}) were considered. Changes in average annual emissions and their monetary values were calculated in the same way as for CO_2, with utility-specific emissions factors for electricity (Ottinger et al. 1990, US EPA 2006a). The price of emissions savings was derived from models that calculate the marginal cost of controlling different pollutants to meet air quality standards (Wang and Santini 1995). Emissions concentrations were obtained from U.S. EPA (2006b) (table 15), and population estimates from the U.S. Census Bureau (2006).

Calculating pollutant uptake by trees—
Trees also remove pollutants from the atmosphere. The modeling method we applied was developed by Scott et al. (1998). It calculates hourly pollutant dry deposition per tree expressed as the product of deposition velocity ($V_d = 1/[R_a + R_b + R_c]$), pollutant concentration (C), canopy-projection area (CP), and a time step, where R_a, R_b, and R_c are aerodynamic, boundary layer, and stomatal resistances. Hourly deposition velocities for each pollutant were calculated during the growing season by using estimates for the resistances ($R_a + R_b + R_c$) for each hour throughout the year. Hourly concentrations for O_3, PM_{10}, and NO_2 for Honolulu, were obtained from the U.S. EPA (2006b), and hourly meteorological data (i.e., air temperature and windspeed) were obtained from the National Climatic Data Center (2006). Solar radiation data were obtained from the Hawaii Institute of Marine Biology (Naughton 2006). The year 2005 was chosen because maximum hourly average ozone and maximum 24-hour average PM_{10} concentrations most closely approximated the average value of those maxima during the last 5-year period. To set a value for pollutant uptake by trees, we used the procedure described above for emissions reductions (table 15). The monetary value for NO_2 was used for ozone.

Estimating biogenic volatile organic compounds emissions from trees—
Annual emissions for biogenic volatile organic compounds (BVOCs) were estimated for the three tree species by using the algorithms of Guenther et al. (1991, 1993). Annual emissions were simulated during the growing season over 40 years. The emission of carbon as isoprene was expressed as a product of the base emission rate (μg C per g dry foliar biomass per hour), adjusted for sunlight and temperature and the amount of dry, foliar biomass present in the tree. Monoterpene emissions were estimated by using a base emission rate adjusted for temperature. The base emission rates for the four species were based on values reported in the literature (Benjamin and Winer 1998). Hourly emissions were summed to get monthly and annual emissions.

Annual dry foliar biomass was derived from field data collected in Honolulu, during September of 2005. The amount of foliar biomass present for each year of the simulated tree's life was unique for each species. Hourly air temperature and solar radiation data for 2005 described in the pollutant uptake section were used as model inputs.

Calculating net air quality benefits—

Net air quality benefits were calculated by subtracting the costs associated with BVOC emissions from benefits associated with pollutant uptake and avoided powerplant emissions. The ozone-reduction benefit from lowering summertime air temperatures, thereby reducing hydrocarbon emissions from **anthropogenic** and **biogenic** sources, was estimated as a function of canopy cover following McPherson and Simpson (1999). They used peak summer air temperature reductions of 0.4 °F for each percentage increase in canopy cover. Hourly changes in air temperature were calculated by reducing this peak air temperature at every hour based on hourly maximum and minimum temperature for that day, scaled by magnitude of maximum total global solar radiation for each day relative to the maximum value for the year.

Stormwater Benefits

Estimating rainfall interception by tree canopies—

A numerical simulation model was used to estimate annual rainfall interception (Xiao et al. 2000). The interception model accounted for water intercepted by the tree, as well as throughfall and **stem flow**. Intercepted water is stored temporarily on canopy leaf and bark surfaces. Rainwater evaporates or drips from leaf surfaces and flows down the stem surface to the ground. Tree-canopy parameters that affect interception include species, leaf and stem surface areas, **shade coefficients** (visual density of the crown), foliation periods, and tree dimensions (e.g., tree height, crown height, crown diameter, and d.b.h.). Tree-height data were used to estimate windspeed at different heights above the ground and resulting rates of evaporation.

The volume of water stored in the tree crown was calculated from crown-projection area (area under tree dripline), **leaf area indices** (LAI, the ratio of LSA to crown projection area), and the depth of water captured by the canopy surface. Gap fractions, foliation periods, and tree surface saturation storage capacity influence the amount of projected throughfall. Tree surface saturation was 1.0 mm (0.04 in) for all trees.

Hourly meteorological and rainfall data for 2005 from the Honolulu International Airport climate monitoring station (National Oceanic and Atmospheric

Administration/National Weather Service, site number: 511919, latitude: 21° 19′ N, longitude: 157°567′ W, elevation: 7 ft) in Honolulu, Hawaii, were used in this simulation. The year 2005 was chosen because it most closely approximated the 10-year average rainfall of 16.2 in (410.7 mm). Annual precipitation in Honolulu during 2005 was 15.4 in (392.4 mm). Storm events less than 0.1 in (2.5 mm) were assumed not to produce runoff and were dropped from the analysis. More complete descriptions of the interception model can be found in Xiao et al. (1998, 2000).

Calculating water quality protection and flood control benefit—
To estimate the value of rainfall intercepted by urban trees, stormwater management control costs were based on construction and operation costs for a typical detention/ retention basin in Honolulu. Twenty-year costs were annualized and divided by the amount of runoff captured in the basin over the course of a typical year. Developers are required to construct detention/retention basins in new projects following local engineering guidelines, which were used in this analysis (City and County of Honolulu 2000). The developed area was 100 acres and the 1-ac basin was designed to hold and treat 58.2 ft^3 of runoff each year (18.9 million gal). The real estate cost for the 1-ac site was $3.75 million, or $187,567 when annualized for a 20-year period (Magota 2007). Constructing the basin was estimated to cost $131,534, or $6,577 annually (US EPA 2000). Operation and maintenance costs were $719 per year. The total average annual cost was $194,863. The average annual control cost was $0.01/gal.

Aesthetic and Other Benefits

Many benefits attributed to urban trees are difficult to translate into economic terms. Beautification, privacy, wildlife habitat, shade that increases human comfort, sense of place and well-being are services that are difficult to price. However, the value of some of these benefits may be captured in the property values of the land on which trees stand.

To estimate the value of these "other" benefits, we applied results of research that compared differences in sales prices of houses to statistically quantify the difference associated with trees. All else being equal, the difference in sales price reflects the willingness of buyers to pay for the benefits and costs associated with trees. This approach has the virtue of capturing in the sales price both the benefits and costs of trees as perceived by the buyers. Limitations to this approach include difficulty determining the value of individual trees on a property, the need to extrapolate results from studies done years ago in other parts of the country, and the need to extrapolate results from front-yard trees on residential properties to trees in other locations (e.g., back yards, streets, parks, and nonresidential land).

Anderson and Cordell (1988) surveyed 844 single-family residences in Athens, Georgia, and found that each large front-yard tree was associated with a 0.88 percent increase in the average home sales price. This percentage of sales price was used as an indicator of the additional value a resident in the Tropical region would gain from selling a home with a large tree.

We averaged the median home prices for the Honolulu ($625,300; National Association of Realtors 2007) and Miami/Ft. Lauderdale/Miami Beach ($345,900; National Association of Realtors 2007) metropolitan statistical areas and for Puerto Rico ($135,000; Freddie Mac 2006) as our starting point ($368,733). Therefore, the value of a large tree that added 0.88-percent to the sales price of such a home was $3,251. To estimate annual benefits, the total added value was divided by the LSA of a 40-year-old monkeypod ($3,251 per 3,670 ft^2) to yield the base value of LSA, $0.96 per ft^2. This value was multiplied by the amount of LSA added to the tree during 1 year of growth.

Additionally, not all street trees are as effective as front-yard trees in increasing property values. For example, trees adjacent to multifamily housing units will not increase the property value at the same rate as trees in front of single-family homes (Gonzales 2004). Therefore, a citywide street tree reduction factor (0.83) was applied to prorate trees' value based on the assumption that trees adjacent to different land uses make different contributions to property sales prices. For this analysis, the street reduction factor reflects the distribution of street trees in Honolulu by land use. Reduction factors were single-home residential (100 percent), multihome residential (70 percent), small commercial (66 percent), industrial/institutional/large commercial (40 percent), park/vacant/other (40 percent) (Gonzales 2004, McPherson 2001).

Calculating the aesthetic and other benefits of residential yard trees—
To calculate the base value for a large tree on private residential property we assumed that a 40-year-old monkeypod in the front yard increased the property sales price by $3,251. Approximately 75 percent of all yard trees, however, are in back yards (Richards et al. 1984). Lacking specific research findings, it was assumed that back-yard trees had 75 percent of the impact on "curb appeal" and sales price compared to front-yard trees. The average annual aesthetic and other benefits for a tree on private property were, therefore, estimated as $0.60 per ft^2 LSA. To estimate annual benefits, this value was multiplied by the amount of LSA added to the tree during 1 year of growth.

Calculating the aesthetic value of a public tree—
The base value of street trees was calculated in the same way as yard trees. However, because street trees may be adjacent to land with little resale potential, an adjusted value was calculated. An analysis of street trees in Modesto, California, sampled from aerial photographs (sample size 8 percent of street trees), found that 15 percent were located adjacent to nonresidential or commercial property (McPherson et al. 1999a). We assumed that 33 percent of these trees—or 5 percent of the entire street-tree population—produced no benefits associated with property value increases.

Although the impact of parks on real estate values has been reported (Hammer et al. 1974, Schroeder 1982, Tyrvainen 1999), to our knowledge, the onsite and external benefits of park trees alone have not been isolated (More et al. 1988). After reviewing the literature and recognizing an absence of data, we made the conservative estimate that park trees had half the impact on property prices of street trees.

Given these assumptions, typical large street and park trees were estimated to increase property values by \$0.74 and \$0.44 per ft^2 LSA, respectively. Assuming that 80 percent of all municipal trees were on streets and 20 percent in parks, a weighted average benefit of \$.68/ft^2 LSA was calculated for each tree.

Calculating Costs

Tree management costs were estimated based on an interview with Honolulu's urban forester (Koike 2006, Oka 2006) and on surveys with municipal foresters in Lantana and Naples, Florida. In addition, several commercial arborists from Hawaii and Florida provided information on tree management costs on residential properties.

Planting

Planting costs include the cost of the tree and the cost for planting, staking, mulching, and establishment irrigation if necessary. Based on our survey of Tropical municipal and commercial arborists, planting costs ranged widely from \$125 for a 5-gal tree to \$2,000 for very large trees. In this analysis we assumed that a 25-gal yard tree was planted at a cost of \$300. The cost for planting a 15- to 25-gal public tree was \$120.

Pruning

Pruning costs for public trees—

After studying data from municipal forestry programs and their contractors, we assumed that young public trees were inspected and pruned once every 2 years during the first 5 years after planting, at a cost of $20 per tree. After this training period, inspection and pruning occurred once every 3 to 4 years. Pruning for small trees (< 20 ft tall) cost $20 per tree. More expensive equipment and more time was required to prune medium ($65 per tree) and large trees ($125 per tree). After factoring in pruning frequency, annualized costs were $10, $6.60, $21.45, and $31.25 per tree for public young, small, medium, and large trees, respectively.

Pruning costs for yard trees—

Based on findings from our survey of commercial arborists in the Tropical region, pruning cycles for yard trees were similar to public trees. Young trees (first 5 years after planting) were pruned every other year, and small, medium, and large trees were pruned every 3 years. Only about 20 percent of all private trees, however, were professionally pruned (contract rate), although the number of professionally pruned trees grows as the trees grow. We assumed that professionals are paid to prune all large trees, 60 percent of the medium trees, and only 6 percent of the small and young trees and conifers (Summit and McPherson 1998). Using these contract rates, along with average pruning prices ($40, $40, $150, and $300 for young, small, medium, and large trees, respectively), the average annual costs for pruning a yard tree were $0.24, $0.16, $5.94, and $19.80 for young, small, medium, and large trees, respectively.

Tree and Stump Removal

The costs for tree removal and disposal were $16 per in d.b.h. for public trees, and $25 per in d.b.h. for yard trees. Stump removal costs were $5 per in d.b.h. for public trees and $8 per in d.b.h. for yard trees. Therefore, total costs for removal and disposal of trees and stumps were $21 per in d.b.h. for public trees, and $33 per in d.b.h. for yard trees. Removal costs of trees under 3 inches in diameter were $20 for yard and public trees.

Other Costs for Public and Yard Trees

Other costs associated with the management of trees include expenditures for infrastructure repair/root pruning, leaf-litter cleanup, and inspection/administration.

Infrastructure conflict costs—

As trees and sidewalks age, roots can cause damage to sidewalks, curbs, paving, and sewer lines. Sidewalk repair is typically one of the largest expenses for public trees (McPherson and Peper 1995). Infrastructure-related expenditures for public trees in Tropical communities were approximately $6 per tree on an annual basis. Roots from most trees in yards do not damage sidewalks and sewers. Therefore, the cost for yard trees was estimated to be only 10 percent of the cost for public trees.

Inspection and administration costs—

Municipal tree programs have administrative costs for salaries of supervisors and clerical staff, operating costs, and overhead. Our survey found that the average annual cost for inspection and administration associated with street- and park-tree management was $1.80 per tree ($0.216 per in d.b.h.). Trees on private property do not accrue this expense.

Calculating Net Benefits

Benefits Accrue at Different Scales

When calculating net benefits, it is important to recognize that trees produce benefits that accrue both on- and offsite. Benefits are realized at four scales: parcel, neighborhood, community, and global. For example, property owners with onsite trees not only benefit from increased property values, but they may also directly benefit from improved human health (e.g., reduced exposure to cancer-causing ultraviolet radiation) and greater psychological well-being through visual and direct contact with plants. However, on the cost side, increased health care costs owing to allergies and respiratory ailments related to pollen may be incurred because of nearby trees. We assume that these intangible benefits and costs are reflected in what we term "aesthetics and other benefits."

The property owner can obtain additional economic benefits from onsite trees depending on their location and condition. For example, carefully located onsite trees can provide air-conditioning savings by shading windows and walls and cooling building microclimates. This benefit can extend to adjacent neighbors who benefit from shade and air-temperature reductions that lower their cooling costs.

Neighborhood attractiveness and property values can be influenced by the extent of tree canopy cover on individual properties. At the community scale, benefits are realized through cleaner air and water, as well as social, educational, and employment and job training benefits that can reduce costs for health care, welfare, crime prevention, and other social service programs.

Reductions in atmospheric CO_2 concentrations owing to trees are an example of benefits that are realized at the global scale.

Annual benefits are calculated as:

$$B = E + AQ + CO_2 + H + A$$

where

E = value of net annual energy savings (cooling and heating)

AQ = value of annual air-quality improvement (pollutant uptake, avoided powerplant emissions, and BVOC emissions)

CO_2 = value of annual CO_2 reductions (sequestration, avoided emissions, release from tree care and decomposition)

H = value of annual stormwater-runoff reductions

A = value of annual aesthetics and other benefits

On the other side of the benefit-cost equation are costs for tree planting and management. Expenditures are borne by property owners (irrigation, pruning, and removal) and the community (pollen and other health care costs). Annual costs (C) are the sum of costs for residential yard trees (C_Y) and public trees (C_P) where:

$$C_Y = P + T + R + D + I + S + Cl + L$$
$$C_P = P + T + R + D + I + S + Cl + L + A$$

where

P = cost of tree and planting

T = average annual tree pruning cost

R = annualized tree and stump removal and disposal cost

D = average annual pest- and disease-control cost

I = annual irrigation cost

S = average annual cost to repair/mitigate infrastructure damage

Cl = annual litter and storm cleanup cost

L = average annual cost for litigation and settlements from tree-related claims

A = annual program administration, inspection, and other costs

Net benefits are calculated as the difference between total benefits and costs:

Net benefits = $B - C$

Benefit-cost ratios (BCR) are calculated as the ratio of benefits to costs:

BCR = B / C

www.ingramcontent.com/pod-product-compliance
Lightning Source LLC
Chambersburg PA
CBHW081500170526
45166CB00008B/2502